T0320658

INTRODUCTION TO MODERN DIGITAL HOLOGRAPHY
With MATLAB®

Get up to speed with digital holography with this concise and straightforward introduction to modern techniques and conventions.

Building up from the basic principles of optics, this book describes key techniques in digital holography, such as phase-shifting holography, low-coherence holography, diffraction tomographic holography, and optical scanning holography. Practical applications are discussed, and accompanied by all the theory necessary to understand the underlying principles at work. A further chapter covers advanced techniques for producing computer-generated holograms. Extensive MATLAB code is integrated with the text throughout and is available for download online, illustrating both theoretical results and practical considerations such as aliasing, zero padding, and sampling.

Accompanied by end-of-chapter problems, and an online solutions manual for instructors, this is an indispensable resource for students, researchers, and engineers in the fields of optical image processing and digital holography.

TING-CHUNG POON is a Professor of Electrical and Computer Engineering at Virginia Tech, and a Visiting Professor at the Shanghai Institute of Optics and Fine Mechanics, Chinese Academy of Sciences. He is a Fellow of the OSA and SPIE.

JUNG-PING LIU is a Professor in the Department of Photonics at Feng Chia University, Taiwan.

INTRODUCTION TO MODERN DIGITAL HOLOGRAPHY

With MATLAB®

TING-CHUNG POON
Virginia Tech, USA

JUNG-PING LIU
Feng Chia University, Taiwan

CAMBRIDGE
UNIVERSITY PRESS

University Printing House, Cambridge CB2 8BS, United Kingdom

One Liberty Plaza, 20th Floor, New York, NY 10006, USA

477 Williamstown Road, Port Melbourne, VIC 3207, Australia

314-321, 3rd Floor, Plot 3, Splendor Forum, Jasola District Centre, New Delhi - 110025, India

79 Anson Road, #06-04/06, Singapore 079906

Cambridge University Press is part of the University of Cambridge.

It furthers the University's mission by disseminating knowledge in the pursuit of education, learning and research at the highest international levels of excellence.

www.cambridge.org
Information on this title: www.cambridge.org/9781107016705

First published 2014
Reprinted with corrections 2014

A catalogue record for this publication is available from the British Library

Library of Congress Cataloging in Publication data
Poon, Ting-Chung.
Introduction to modern digital holography : with MATLAB / Ting-Chung Poon, Jung-Ping Liu.
pages cm
ISBN 978-1-107-01670-5 (Hardback)
1. Holography–Data processing. 2. Image processing–Digital techniques. I. Liu, Jung-Ping. II. Title.
TA1542.P66 2014
621.36′75–dc23
2013036072

ISBN 978-1-107-01670-5 Hardback

Additional resources for this publication at www.cambridge.org/digitalholography

Contents

Preface

Owing to the advance in faster electronics and digital processing power, the past decade has seen an impressive re-emergence of digital holography. Digital holography is a topic of growing interest and it finds applications in three-dimensional imaging, three-dimensional displays and systems, as well as biomedical imaging and metrology. While research in digital holography continues to be vibrant and digital holography is maturing, we find that there is a lack of textbooks in the area. The present book tries to serve this need: to promote and teach the foundations of digital holography. In addition to presenting traditional digital holography and applications in Chapters 1–4, we also discuss modern applications and techniques in digital holography such as phase-shifting holography, low-coherence holography, diffraction tomographic holography, optical scanning holography, sectioning in holography, digital holographic microscopy as well as computer-generated holography in Chapters 5–7. This book is geared towards undergraduate seniors or first-year graduate-level students in engineering and physics. The material covered is suitable for a one-semester course in Fourier optics and digital holography. The book is also useful for scientists and engineers, and for those who simply want to learn about optical image processing and digital holography.

We believe in the inclusion of MATLAB® in the textbook because digital holography relies heavily on digital computations to process holographic data. MATLAB® will help the reader grasp and visualize some of the important concepts in digital holography. The use of MATLAB® not only helps to illustrate the theoretical results, but also makes us aware of computational issues such as aliasing, zero padding, sampling, etc. that we face in implementing them. Nevertheless, this text is not about teaching MATLAB®, and some familiarity with MATLAB® is required to understand the codes.

The MATLAB® codes included in this book are all available to download from the publisher at www.cambridge.org/digitalholography.

Ting-Chung Poon would like to thank his wife, Eliza, and his children, Christina and Justine, for their love. This year is particularly special to him as Christina gave birth to a precious little one – Gussie. Jung-Ping Liu would like to thank his wife, Hui-Chu, and his parents for their understanding and encouragement.

1

Wave optics

1.1 Maxwell's equations and the wave equation

In wave optics, we treat light as waves. Wave optics accounts for wave effects such as interference and diffraction. The starting point for wave optics is Maxwell's equations:

$$\nabla \cdot \boldsymbol{D} = \rho_v, \tag{1.1}$$

$$\nabla \cdot \boldsymbol{B} = 0, \tag{1.2}$$

$$\nabla \times \boldsymbol{E} = -\frac{\partial \boldsymbol{B}}{\partial t}, \tag{1.3}$$

$$\nabla \times \boldsymbol{H} = \boldsymbol{J} = \boldsymbol{J}_C + \frac{\partial \boldsymbol{D}}{\partial t}, \tag{1.4}$$

where we have four vector quantities called electromagnetic (EM) fields: the electric field strength $\boldsymbol{E}$ (V/m), the electric flux density $\boldsymbol{D}$ $\left(\text{C/m}^2\right)$, the magnetic field strength $\boldsymbol{H}$ (A/m), and the magnetic flux density $\boldsymbol{B}$ $\left(\text{Wb/m}^2\right)$. The vector quantity $\boldsymbol{J}_C$ and the scalar quantity ρ_v are the current density $\left(\text{A/m}^2\right)$ and the electric charge density $\left(\text{C/m}^3\right)$, respectively, and they are the sources responsible for generating the electromagnetic fields. In order to determine the four field quantities completely, we also need the constitutive relations

$$\boldsymbol{D} = \varepsilon \boldsymbol{E}, \tag{1.5}$$

and

$$\boldsymbol{B} = \mu \boldsymbol{H}, \tag{1.6}$$

where ε and μ are the permittivity (F/m) and permeability (H/m) of the medium, respectively. In the case of a linear, homogenous, and isotropic medium such as in vacuum or free space, ε and μ are scalar constants. Using Eqs. (1.1)–(1.6), we can derive a wave equation in $\boldsymbol{E}$ or $\boldsymbol{B}$ in free space. For example, by taking the curl of $\boldsymbol{E}$ in Eq. (1.3), we can derive the wave equation in $\boldsymbol{E}$ as

$$\nabla^2\boldsymbol{E} - \mu\varepsilon\frac{\partial^2\boldsymbol{E}}{\partial t^2} = \mu\frac{\partial\boldsymbol{J}_C}{\partial t} + \frac{1}{\varepsilon}\nabla\rho_v, \tag{1.7}$$

where $\nabla^2 = \partial^2/\partial x^2 + \partial^2/\partial y^2 + \partial^2/\partial z^2$ is the Laplacian operator in Cartesian coordinates. For a source-free medium, i.e., $\boldsymbol{J}_C = 0$ and $\rho_v = 0$, Eq. (1.7) reduces to the homogeneous wave equation:

$$\nabla^2\boldsymbol{E} - \frac{1}{v^2}\frac{\partial^2\boldsymbol{E}}{\partial t^2} = 0. \tag{1.8}$$

Note that $v = 1/\sqrt{\mu\varepsilon}$ is the velocity of the wave in the medium. Equation (1.8) is equivalent to three scalar equations, one for every component of $\boldsymbol{E}$. Let

$$\boldsymbol{E} = E_x\boldsymbol{a}_x + E_y\boldsymbol{a}_y + E_z\boldsymbol{a}_z, \tag{1.9}$$

where $\boldsymbol{a_x}$, $\boldsymbol{a_y}$, and $\boldsymbol{a_z}$ are the unit vectors in the x, y, and z directions, respectively. Equation (1.8) then becomes

$$\left(\frac{\partial^2}{\partial x^2} + \frac{\partial^2}{\partial y^2} + \frac{\partial^2}{\partial z^2}\right)(E_x\boldsymbol{a_x} + E_y\boldsymbol{a_y} + E_z\boldsymbol{a_z}) = \frac{1}{v^2}\frac{\partial^2}{\partial t^2}(E_x\boldsymbol{a_x} + E_y\boldsymbol{a_y} + E_z\boldsymbol{a_z}). \tag{1.10}$$

Comparing the $\boldsymbol{a_x}$-component on both sides of the above equation gives us

$$\frac{\partial^2 E_x}{\partial x^2} + \frac{\partial^2 E_x}{\partial y^2} + \frac{\partial^2 E_x}{\partial z^2} = \frac{1}{v^2}\frac{\partial^2 E_x}{\partial t^2}.$$

Similarly, we can derive the same type of equation shown above for the E_y and E_z components by comparison with other components in Eq. (1.10). Hence we can write a compact equation for the three components as

$$\frac{\partial^2\psi}{\partial x^2} + \frac{\partial^2\psi}{\partial y^2} + \frac{\partial^2\psi}{\partial z^2} = \frac{1}{v^2}\frac{\partial^2\psi}{\partial t^2} \tag{1.11a}$$

or

$$\nabla^2\psi = \frac{1}{v^2}\frac{\partial^2\psi}{\partial t^2}, \tag{1.11b}$$

where ψ can represent a component, E_x, E_y, or E_z, of the electric field $\boldsymbol{E}$. Equation (1.11) is called the *three-dimensional scalar wave equation*. We shall look at some of its simplest solutions in the next section.

1.2 Plane waves and spherical waves

In this section, we will examine some of the simplest solutions, namely the plane wave solution and the spherical wave solution, of the three-dimensional scalar wave equation in Eq. (1.11). For simple harmonic oscillation at angular frequency ω_0 (radian/second) of the wave, in Cartesian coordinates, the plane wave solution is

$$\psi(x, y, z, t) = A\exp[j(\omega_0 t - \boldsymbol{k}_0 \cdot \boldsymbol{R})], \tag{1.12}$$

where $j = \sqrt{-1}$, $\boldsymbol{k}_0 = k_{0x}\boldsymbol{a}_x + k_{0y}\boldsymbol{a}_y + k_{0z}\boldsymbol{a}_z$ is the propagation vector, and $\boldsymbol{R} = x\boldsymbol{a}_x + y\boldsymbol{a}_y + z\boldsymbol{a}_z$ is the position vector. The magnitude of $\boldsymbol{k}_0$ is called the wave number and is $|\boldsymbol{k}_0| = k_0 = \sqrt{k_{0x}^2 + k_{0y}^2 + k_{0z}^2} = \omega_0/v$. If the medium is free space, $v = c$ (the speed of light in vacuum) and k_0 becomes the wave number in free space. Equation (1.12) is a *plane wave* of amplitude A, traveling along the $\boldsymbol{k}_0$ direction. The situation is shown in Fig. 1.1.

If a plane wave is propagating along the positive z-direction, Eq. (1.12) becomes

$$\psi(z, t) = A\exp[j(\omega_0 t - k_0 z)], \tag{1.13}$$

which is a solution to the one-dimensional scalar wave equation given by

$$\frac{\partial^2 \psi}{\partial z^2} = \frac{1}{v^2}\frac{\partial^2 \psi}{\partial t^2}. \tag{1.14}$$

Equation (1.13) is a complex representation of a plane wave. Since the electromagnetic fields are real functions of space and time, we can represent the plane wave in real quantities by taking the real part of ψ to obtain

$$\mathrm{Re}\{\psi(z, t)\} = A\cos[(\omega_0 t - k_0 z)]. \tag{1.15}$$

Another important solution to the wave equation in Eq. (1.11) is a spherical wave solution. The spherical wave solution is a solution which has spherical symmetry, i.e., the solution is not a function of ϕ and θ under the spherical coordinates shown in Fig. 1.2. The expression for the Laplacian operator, ∇^2, is

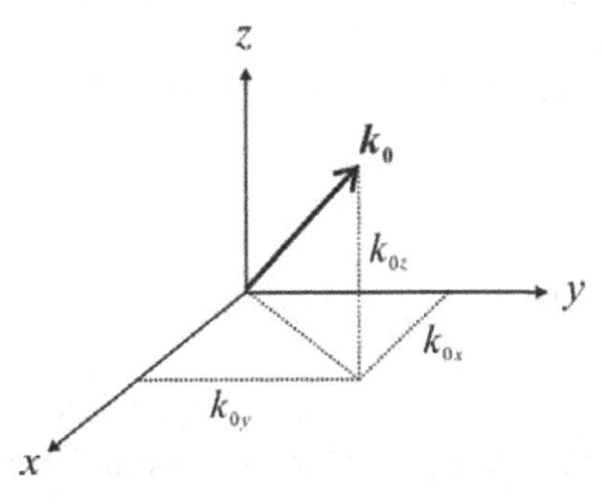

Figure 1.1 Plane wave propagating along the direction $\boldsymbol{k}_0$.

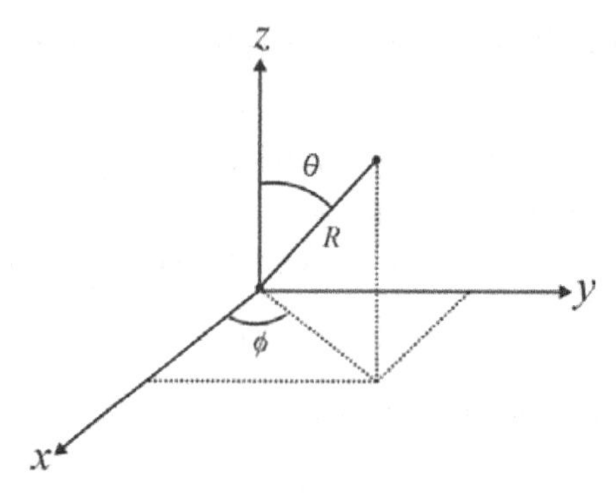

Figure 1.2 Spherical coordinate system.

$$\nabla^2 = \frac{\partial^2}{\partial R^2} + \frac{2}{R}\frac{\partial}{\partial R} + \frac{1}{R^2 \sin^2\theta}\frac{\partial^2}{\partial \phi^2} + \frac{1}{R^2}\frac{\partial^2}{\partial \theta^2} + \frac{\cot\theta}{R^2}\frac{\partial}{\partial \theta}.$$

Hence Eq. (1.11b), under spherical symmetry, becomes

$$\frac{\partial^2 \psi}{\partial R^2} + \frac{2}{R}\frac{\partial \psi}{\partial R} = \frac{1}{v^2}\frac{\partial^2 \psi}{\partial t^2}. \tag{1.16}$$

Since

$$R\left(\frac{\partial^2 \psi}{\partial R^2} + \frac{2}{R}\frac{\partial \psi}{\partial R}\right) = \frac{\partial^2 (R\psi)}{\partial R^2},$$

we can re-write Eq. (1.16) to become

$$\frac{\partial^2 (R\psi)}{\partial R^2} = \frac{1}{v^2}\frac{\partial^2 (R\psi)}{\partial t^2}. \tag{1.17}$$

By comparing the above equation with Eq. (1.14), which has a solution given by Eq. (1.13), we can construct a simple solution to Eq. (1.17) as

$$R\psi(R,t) = A\exp[j(\omega_0 t - k_0 R)],$$

or

$$\psi(R,t) = \frac{A}{R}\exp[j(\omega_0 t - k_0 R)]. \tag{1.18}$$

The above equation is a *spherical wave* of amplitude A, which is one of the solutions to Eq. (1.16). In summary, plane waves and spherical waves are some of the simplest solutions of the three-dimensional scalar wave equation.

1.3 Scalar diffraction theory

For a plane wave incident on an aperture or a diffracting screen, i.e., an opaque screen with some openings allowing light to pass through, we need to find the field distribution exiting the aperture or the diffracted field. To tackle the diffraction problem, we find the solution of the scalar wave equation under some initial condition. Let us assume the aperture is represented by a transparency with *amplitude transmittance*, often called *transparency function*, given by $t(x, y)$, located on the plane $z = 0$ as shown in Fig. 1.3.

A plane wave of amplitude A is incident on the aperture. Hence at $z = 0$, according to Eq. (1.13), the plane wave immediately in front of the aperture is given by $A \exp(j\omega_0 t)$. The field distribution immediately after the aperture is $\psi(x, y, z = 0, t) = At(x, y) \exp(j\omega_0 t)$. In general, $t(x, y)$ is a complex function that modifies the field distribution incident on the aperture, and the transparency has been assumed to be infinitely thin. To develop $\psi(x, y, z = 0, t)$ further mathematically, we write

$$\psi(x, y, z = 0, t) = At(x, y)\exp(j\omega_0 t) = \psi_p(x, y; z = 0)\exp(j\omega_0 t)$$
$$= \psi_{p0}(x, y)\exp(j\omega_0 t). \tag{1.19}$$

The quantity $\psi_{p0}(x, y)$ is called the *complex amplitude* in optics. This complex amplitude is the initial condition, which is given by $\psi_{p0}(x, y) = A \times t(x, y)$, the amplitude of the incident plane wave multiplied by the transparency function of the aperture. To find the field distribution at z away from the aperture, we model the solution in the form of

$$\psi(x, y, z, t) = \psi_p(x, y; z)\exp(j\omega_0 t), \tag{1.20}$$

where $\psi_p(x, y; z)$ is the unknown to be found with initial condition $\psi_{p0}(x, y)$ given. To find $\psi_p(x, y; z)$, we substitute Eq. (1.20) into the three-dimensional scalar wave equation given by Eq. (1.11a) to obtain the *Helmholtz equation* for $\psi_p(x, y; z)$,

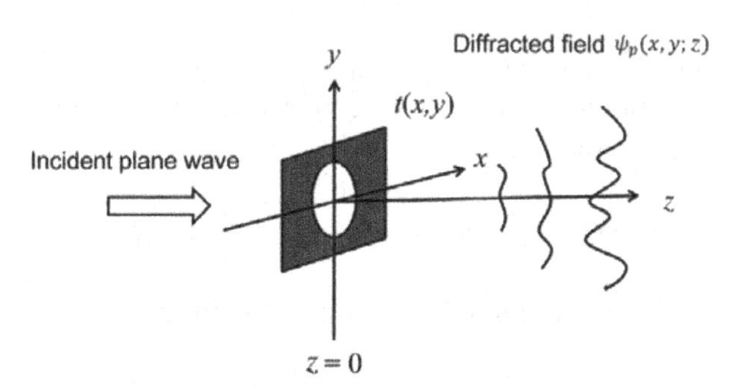

Figure 1.3 Diffraction geometry: $t(x, y)$ is a diffracting screen.

$$\frac{\partial^2 \psi_p}{\partial x^2} + \frac{\partial^2 \psi_p}{\partial y^2} + \frac{\partial^2 \psi_p}{\partial z^2} + k_0^2 \psi_p = 0. \tag{1.21}$$

To find the solution to the above equation, we choose to use the Fourier transform technique. The two-dimensional Fourier transform of a spatial signal $f(x, y)$ is defined as

$$\mathcal{F}\{f(x,y)\} = F(k_x, k_y) = \iint_{-\infty}^{\infty} f(x,y) \exp(jk_x x + jk_y y) dx\, dy, \tag{1.22a}$$

and the inverse Fourier transform is

$$\mathcal{F}^{-1}\{F(k_x, k_y)\} = f(x,y) = \frac{1}{4\pi^2} \iint_{-\infty}^{\infty} F(k_x, k_y) \exp(-jk_x x - jk_y y) dk_x\, dk_y, \tag{1.22b}$$

where k_x and k_y are called spatial radian frequencies as they have units of radian per unit length. The functions $f(x, y)$ and $F(k_x, k_y)$ form a Fourier transform pair. Table 1.1 shows some of the most important transform pairs.

By taking the two-dimensional Fourier transform of Eq. (1.21) and using transform pair number 4 in Table 1.1 to obtain

$$\begin{aligned} \mathcal{F}\left\{\frac{\partial^2 \psi_p}{\partial x^2}\right\} &= (-jk_x)^2 \Psi_\mathrm{p}(k_x, k_y; z) \\ \mathcal{F}\left\{\frac{\partial^2 \psi_p}{\partial y^2}\right\} &= (-jk_y)^2 \Psi_\mathrm{p}(k_x, k_y; z), \end{aligned} \tag{1.23}$$

where $\mathcal{F}\{\psi_p(x, y; z)\} = \Psi_\mathrm{p}(k_x, k_y; z)$, we have a differential equation in $\Psi_\mathrm{p}(k_x, k_y; z)$ given by

$$\frac{d^2 \Psi_\mathrm{p}}{dz^2} + k_0^2 \left(1 - \frac{k_x^2}{k_0^2} - \frac{k_y^2}{k_0^2}\right) \Psi_\mathrm{p} = 0 \tag{1.24}$$

subject to the initial known condition $\mathcal{F}\{\psi_p(x, y; z = 0)\} = \Psi_\mathrm{p}(k_x, k_y; z = 0) = \Psi_\mathrm{p0}(k_x, k_y)$. The solution to the above second ordinary differential equation is straightforward and is given by

$$\Psi_\mathrm{p}(k_x, k_y; z) = \Psi_\mathrm{p0}(k_x, k_y) \exp\left[-jk_0 \sqrt{(1 - k_x^2/k_0^2 - k_y^2/k_0^2)} z\right] \tag{1.25}$$

as we recognize that the differential equation of the form

$$\frac{d^2 y(z)}{dz^2} + \alpha^2 y(z) = 0$$

Table 1.1 *Fourier transform pairs*

1. $f(x, y)$	$F(k_x, k_y)$
2. Shifting $f(x - x_0, y - y_0)$	$F(k_x, k_y)\exp(jk_x x_0 + jk_y y_0)$
3. Scaling $f(ax, by)$	$\frac{1}{\lvert ab \rvert} f\left(\frac{k_x}{a}, \frac{k_y}{b}\right)$
4. Differentiation $\partial f(x, y) / \partial x$	$-jk_x F(k_x, k_y)$
5. Convolution integral $f_1 * f_2 = \iint_{-\infty}^{\infty} f_1(x', y') f_2(x-x', y-y') dx' dy'$	Product of spectra $F_1(k_x, k_y) F_2(k_x, k_y)$ where $\mathcal{F}\{f_1(x, y)\} = F_1(k_x, k_y)$ and $\mathcal{F}\{f_2(x, y)\} = F_2(k_x, k_y)$
6. Correlation $f_1 \otimes f_2 = \iint_{-\infty}^{\infty} f_1^*(x', y') f_2(x+x', y+y') dx' dy'$	$F_1^*(k_x, k_y) F_2(k_x, k_y)$
7. Gaussian function $\exp[-\alpha(x^2 + y^2)]$	Gaussian function $\frac{\pi}{\alpha}\exp\left[-\frac{(k_x^2 + k_y^2)}{4\alpha}\right]$
8. Constant of unity 1	Delta function $4\pi^2 \delta(x, y) = \iint_{-\infty}^{\infty} 1 \exp(\pm jk_x x \pm jk_y y) dk_x dk_y$
9. Delta function $\delta(x, y)$	Constant of unity 1
10. Triangular function $\Lambda\left(\frac{x}{a}, \frac{y}{b}\right) = \Lambda\left(\frac{x}{a}\right)\Lambda\left(\frac{y}{b}\right)$, where $\Lambda\left(\frac{x}{a}\right) = \begin{cases} 1 - \left\lvert \frac{x}{a} \right\rvert & \text{for } \left\lvert \frac{x}{a} \right\rvert \leq 1 \\ 0 & \text{otherwise} \end{cases}$	$a \ \mathrm{sinc}^2\left(\frac{k_x a}{2\pi}\right) b \ \mathrm{sinc}^2\left(\frac{k_y b}{2\pi}\right)$
11. Rectangular function $\mathrm{rect}(x, y) = \mathrm{rect}(x)\mathrm{rect}(y)$, where $\mathrm{rect}(x) = \begin{cases} 1 & \text{for } \lvert x \rvert < 1/2 \\ 0 & \text{otherwise} \end{cases}$	Sinc function $\mathrm{sinc}\left(\frac{k_x}{2\pi}, \frac{k_y}{2\pi}\right) = \mathrm{sinc}\left(\frac{k_x}{2\pi}\right)\mathrm{sinc}\left(\frac{k_y}{2\pi}\right)$, where $\mathrm{sinc}(x) = \sin(\pi x)/\pi x$
12. Linear phase (plane wave) $\exp[-j(ax + by)]$	$4\pi^2 \delta(k_x - a, k_y - b)$
13. Quadratic phase (complex Fresnel zone plate CFZP) $\exp[-ja(x^2 + y^2)]$	Quadratic phase (complex Fresnel zone plate CFZP) $\frac{-j\pi}{a}\exp\left[\frac{j}{4a}(k_x^2 + k_y^2)\right]$

has the solution given by

$$y(z) = y(0)\exp[-j\alpha z].$$

From Eq. (1.25), we define the *spatial frequency transfer function of propagation* through a distance z as [1]

$$\begin{aligned}\mathcal{H}(k_x, k_y; z) &= \Psi_\mathrm{p}(k_x, k_y; z)/\Psi_{\mathrm{p}0}(k_x, k_y) \\ &= \exp\left[-jk_0\sqrt{\left(1 - k_x^2/k_0^2 - k_y^2/k_0^2\right)}\, z\right].\end{aligned} \tag{1.26}$$

Hence the complex amplitude $\psi_p(x, y; z)$ is given by the inverse Fourier transform of Eq. (1.25):

$$\begin{aligned}\psi_p(x,y;z) &= \mathcal{F}^{-1}\{\Psi_\mathrm{p}(k_x,k_y;z)\} = \mathcal{F}^{-1}\{\Psi_{\mathrm{p}0}(k_x,k_y)\mathcal{H}(k_x,k_y;z)\} \\ &= \frac{1}{4\pi^2}\iint_{-\infty}^{\infty}\Psi_{\mathrm{p}0}(k_x,k_y)\exp\left[-jk_0\sqrt{\left(1-k_x^2/k_0^2-k_y^2/k_0^2\right)}\, z\right]\exp(-jk_x x-jk_y y)dk_x\, dk_y.\end{aligned} \tag{1.27}$$

The above equation is a very important result. For a given field distribution along the $z = 0$ plane, i.e., $\psi_p\ (x, y; z = 0) = \psi_{p0}(x, y)$, we can find the field distribution across a plane parallel to the (x, y) plane but at a distance z from it by calculating Eq. (1.27). The term $\Psi_{\mathrm{p}0}(k_x, k_y)$ is a Fourier transform of $\psi_{p0}(x, y)$ according to Eq. (1.22):

$$\psi_{p0}(x,y) = \mathcal{F}^{-1}\{\Psi_{\mathrm{p}0}(k_x,k_y)\} = \frac{1}{4\pi^2}\iint_{-\infty}^{\infty}\Psi_{\mathrm{p}0}(k_x,k_y)\exp(-jk_x x-jk_y y)dk_x\, dk_y. \tag{1.28}$$

The physical meaning of the above integral is that we first recognize a plane wave propagating with propagation vector $\boldsymbol{k}_0$, as illustrated in Fig. 1.1. The complex amplitude of the plane wave, according to Eq. (1.12), is given by

$$A\exp(-jk_{0x}x - jk_{0y}y - jk_{0z}z). \tag{1.29}$$

The field distribution at $z = 0$ or the plane wave component with amplitude A is given by

$$A\exp(-jk_{0x}x - jk_{0y}y).$$

Comparing the above equation with Eq. (1.28) and recognizing that the spatial radian frequency variables k_x and k_y of the field distribution $\psi_{p0}(x, y)$ are k_{0x} and k_{0y} of the plane wave in Eq. (1.29), $\Psi_{\mathrm{p}0}(k_x, k_y)$ is called the *angular plane wave spectrum* of the field distribution $\psi_{p0}(x, y)$. Therefore, $\Psi_{\mathrm{p}0}(k_x, k_y)\exp(-jk_x x - jk_y y)$ is the plane wave component with amplitude $\Psi_{\mathrm{p}0}(k_x, k_y)$ and by summing over various directions of k_x and k_y, we have the field distrition $\psi_{p0}(x, y)$ at $z = 0$ given by Eq. (1.28). To find the field distribution a distance of z away, we simply let the various plane wave components propagate over a distance z, which means acquiring a phase shift of $\exp(-jk_z z)$ or $\exp(-jk_{0z}z)$ by noting that the variable k_z is k_{0z} of the plane wave so that we have

$$\begin{aligned}\psi_p(x,y;z) &= \frac{1}{4\pi^2}\iint_{-\infty}^{\infty}\Psi_{p0}(k_x,k_y)\exp(-jk_x x - jk_y y - jk_z z)dk_x\, dk_y \\ &= \mathcal{F}^{-1}\{\Psi_{p0}(k_x,k_y)\exp(-jk_{0z}z)\}.\end{aligned} \tag{1.30}$$

Note that $k_0 = \sqrt{k_{0x}^2 + k_{0y}^2 + k_{0z}^2}$ and hence $k_z = k_{0z} = \pm k_0\sqrt{(1-k_x^2/k_0^2-k_y^2/k_0^2)}$, we immediately recover Eq. (1.27) from Eq. (1.30) and provide physical meaning to the equation. Note that we have kept the + sign in the above relation to represent waves traveling in the positive z-direction. In addition, for propagation of plane waves, $1-k_x^2/k_0^2-k_y^2/k_0^2 \geq 0$ or $k_x^2+k_y^2 \leq k_0^2$. If the reverse is true, i.e., $k_x^2+k_y^2 \geq k_0^2$, we have evanescent waves.

1.3.1 Fresnel diffraction

When propagating waves make small angles, i.e., under the so-called *paraxial approximation*, we have $k_x^2 + k_y^2 \ll k_0^2$ and

$$\sqrt{\left(1-k_x^2/k_0^2-k_y^2/k_0^2\right)} \approx 1 - k_x^2/2k_0^2 - k_y^2/2k_0^2. \tag{1.31}$$

Equation (1.27) becomes

$$\psi_p(x,y;z) = \frac{1}{4\pi^2}\iint_{-\infty}^{\infty}\Psi_{p0}(k_x,k_y)\exp\left[-jk_0 z + j(k_x^2+k_y^2)z/2k_0\right] \times \exp(-jk_x x - jk_y y)dk_x\, dk_y,$$

which can be written in a compact form as

$$\psi_p(x,y;z) = \mathcal{F}^{-1}\{\Psi_{p0}(k_x,k_y)H(k_x,k_y;z)\}, \tag{1.32}$$

where

$$H(k_x,k_y;z) = \exp[-jk_0 z]\exp\left[j(k_x^2+k_y^2)z/2k_0\right]. \tag{1.33}$$

$H(k_x, k_y; z)$ is called the *spatial frequency transfer function in Fourier optics* [1]. The transfer function is simply a paraxial approximation to $\mathcal{H}(k_x,k_y;z)$. The inverse Fourier transform of $H(k_x, k_y; z)$ is known as the *spatial impulse response in Fourier optics*, $h(x, y; z)$ [1]:

$$h(x, y; z) = \mathcal{F}^{-1}\{H(k_x, k_y; z)\} = \exp(-jk_0 z)\frac{jk_0}{2\pi z}\exp\left[\frac{-jk_0}{2z}(x^2 + y^2)\right]. \quad (1.34)$$

To find the inverse transform of the above equation, we have used transform pair number 13 in Table 1.1. We can express Eq. (1.32) in terms of the convolution integral by using transform pair number 5:

$$\begin{aligned}\psi_p(x, y; z) &= \psi_{p0}(x, y) * h(x, y; z) \\ &= \exp(-jk_0 z)\frac{jk_0}{2\pi z}\iint_{-\infty}^{\infty} \psi_{p0}(x', y')\exp\left\{\frac{-jk_0}{2z}\left[(x - x')^2 + (y - y')^2\right]\right\}dx'dy'.\end{aligned} \quad (1.35)$$

Equation (1.35) is called the *Fresnel diffraction formula* and describes the Fresnel diffraction of a "beam" during propagation which has an initial complex amplitude given by $\psi_{p0}(x, y)$.

1.3.2 Fraunhofer diffraction

If we wish to calculate the diffraction pattern at a distance far away from the aperture, Eq. (1.35) can be simplified. To see how, let us complete the square in the exponential function and then re-write Eq. (1.35) as

$$\begin{aligned}\psi_p(x, y; z) = &\exp(-jk_0 z)\frac{jk_0}{2\pi z}\exp\left[\frac{-jk_0}{2z}(x^2 + y^2)\right] \\ &\times \iint_{-\infty}^{\infty} \psi_{p0}(x', y')\exp\left\{\frac{-jk_0}{2z}\left[(x')^2 + (y')^2\right]\right\}\exp\left[\frac{jk_0}{z}(xx' + yy')\right]dx'dy'.\end{aligned} \quad (1.36)$$

In terms of Fourier transform, we can write the Fresnel diffraction formula as follows:

$$\begin{aligned}\psi_p(x, y; z) = &\exp(-jk_0 z)\frac{jk_0}{2\pi z}\exp\left[\frac{-jk_0}{2z}(x^2 + y^2)\right] \\ &\times \mathcal{F}\left\{\psi_{p0}(x, y)\exp\left[\frac{-jk_0}{2z}(x^2 + y^2)\right]\right\}\Bigg|_{k_x = \frac{k_0 x}{z},\, k_y = \frac{k_0 y}{z}}.\end{aligned} \quad (1.37)$$

In the integral shown in Eq. (1.36), ψ_{p0} is considered the "source," and therefore the coordinates x' and y' can be called the source plane. In order to find the field distribution ψ_p on the observation plane z away, or on the x–y plane, we need to multiply the source by the two exponential functions as shown inside the integrand of Eq. (1.36) and then to integrate over the source coordinates. The result of the integration is then multiplied by the factor $\exp(-jk_0z)$ $(jk_0/2\pi z)$ $\exp[(-jk_0/2z)$ $(x^2 + y^2)]$ to arrive at the final result on the observation plane given by Eq. (1.36).

Note that the integral in Eq. (1.36) can be simplified if the approximation below is true:

$$\frac{k_0}{2}\left[(x')^2 + (y')^2\right]_{max} = \frac{\pi}{\lambda_0}\left[(x')^2 + (y')^2\right]_{max} \ll z. \tag{1.38}$$

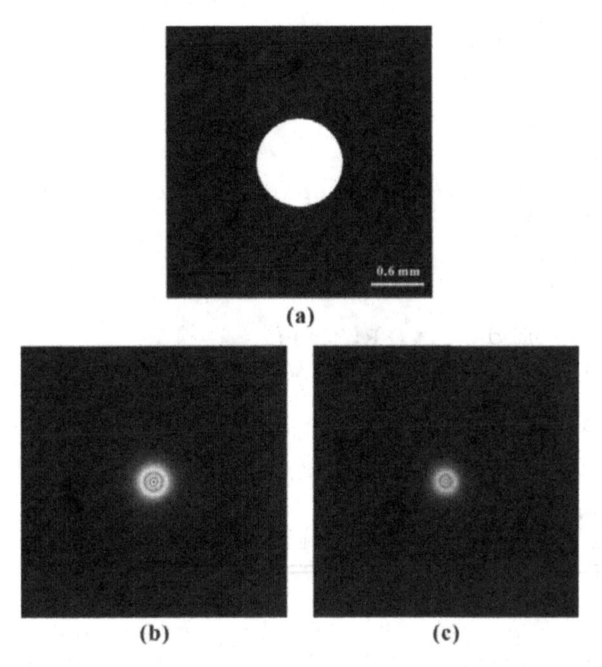

Figure 1.4 (a) $t(x,y)$ is a diffracting screen in the form of circ(r / r_0), $r_0 = 0.5$ mm. (b) Fresnel diffraction at $z = 7$ cm, $|\psi_p(x, y; z = 7 \text{ cm})|$. (c) Fresnel diffraction at $z = 9$ cm, $|\psi_p(x, y; z = 9 \text{ cm})|$. See Table 1.2 for the MATLAB code.

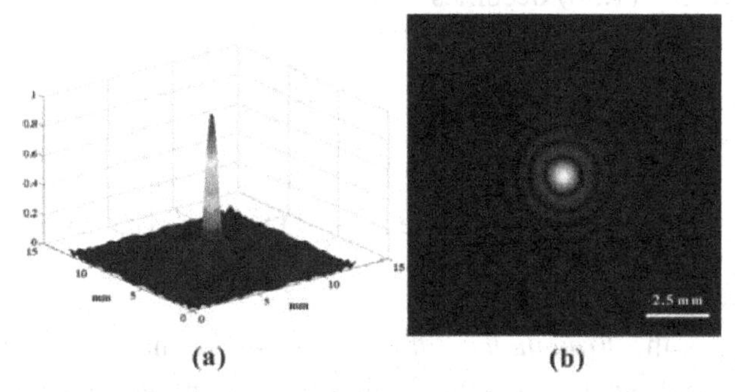

Figure 1.5 (a) Three-dimensional plot of a Fraunhofer diffraction pattern at $z = 1$ m, $|\psi_p(x, y; z = 1 \text{ m})|$. (b) Gray-scale plot of $|\psi_p(x, y; z = 1 \text{ m})|$. See Table 1.3 for the MATLAB code.

Table 1.2 *MATLAB code for Fresnel diffraction of a circular aperture, see Fig. 1.4*

```
close all; clear all;
lambda=0.6*10^-6; % wavelength, unit:m
delta=10*lambda;  % sampling period, unit:m
z=0.07;           % propagation distance; m
M=512;            % space size
c=1:M;
r=1:M;
[C, R]=meshgrid(c, r);
THOR=((R-M/2-1).^2+(C-M/2-1).^2).^0.5;
RR=THOR.*delta;
OB=zeros(M);      % Object
for a=1:M;
   for b=1:M;
      if RR(a,b)<=5*10^-4; % aperture radius unit:m
         OB(a,b)=1;
      end
   end
end
QP=exp(1i*pi/lambda/z.*(RR.^2));
FD=fftshift(fft2(fftshift(OB.*QP)));
FD=abs(FD);
FD=FD/max(max(FD));
figure; imshow(OB);
title('Circular aperture')
figure; imshow(FD);
title('Modulus of the Fresnel diffraction pattern')
```

The term $\pi[(x')^2 + (y')^2]_{max}$ is like the maximum area of the source and if this area divided by the wavelength is much less than the distance z under consideration, the term $\exp\{(-jk_0/2z)[(x')^2 + (y')^2]\}$ inside the integrand can be considered to be unity, and hence Eq. (1.36) becomes

$$\psi_p(x, y; z) = \exp(-jk_0 z)\frac{jk_0}{2\pi z}\exp\left[\frac{-jk_0}{2z}(x^2 + y^2)\right] \times \iint_{-\infty}^{\infty} \psi_{p0}(x', y')\exp\left[\frac{jk_0}{z}(xx' + yy')\right]dx'\,dy'. \tag{1.39}$$

Equation (1.39) is the *Fraunhofer diffraction formula* and is the limiting case of Fresnel diffraction. Equation (1.39) describes the *Fraunhofer approximation* or the far field approximation as diffraction is observed at a far distance. In terms of Fourier transform, we can write the Fraunhofer diffraction formula as follows:

Table 1.3 *MATLAB code for Fraunhofer diffraction of a circular aperture, see Fig. 1.5*

```
close all; clear all;
lambda=0.6*10^-6; %wavelength, unit:m
delta=80*lambda;  %sampling period, unit:m
z=1;              %propagation distance, unit:m
M=512;            %space size
c=1:M;
r=1:M;
[C, R]=meshgrid(c, r);
THOR=(((R-M/2-1).^2+(C-M/2-1).^2).^0.5)*delta;
OB=zeros(M);      % Object
for a=1:M;
   for b=1:M;
      if THOR(a,b)<=5*10^-4; %aperture radius unit:m
         OB(a,b)=1;
      end
   end
end
FD=fftshift(fft2(fftshift(OB)));
FD=abs(FD);
FD=FD/max(max(FD));
C=C*lambda*z/M/delta*1000;
R=R*lambda*z/M/delta*1000;
figure; mesh(R, C, FD);
figure; imshow(FD);
title('Modulus of the Fraunhofer diffraction pattern')
```

$$\psi_p(x, y; z) = \exp(-jk_0 z)\frac{jk_0}{2\pi z}\exp\left[\frac{-jk_0}{2z}\left(x^2 + y^2\right)\right] \times \mathcal{F}\{\psi_{p0}(x, y)\}_{k_x=\frac{k_0 x}{z}, k_y=\frac{k_0 y}{z}}. \tag{1.40}$$

Figure 1.4 shows the simulation of Fresnel diffraction of a circular aperture function circ (r/r_0), i.e., $\psi_{p0}(x, y) = \mathrm{circ}(r/r_0)$, where $r = \sqrt{x^2 + y^2}$ and $\mathrm{circ}(r/r_0)$ denotes a value 1 within a circle of radius r_0 and 0 otherwise. The wavelength used for the simulation is 0.6 μm. Since $\psi_p(x, y; z)$ is a complex function, we plot its absolute value in the figures. Physically, the situation corresponds to the incidence of a plane wave with unit amplitude on an opaque screen with a circular opening with radius r_0 as $\psi_p(x, y; z) = 1 \times t(x, y)$ with $t(x, y) = \mathrm{circ}(r/r_0)$. We would then observe the intensity pattern, which is proportional to $|\psi_p(x, y; z)|^2$, at distance z away from the aperture. In Fig. 1.5, we show Fraunhofer diffraction. We have chosen the distance of 1 m so that the Fraunhofer approximation from Eq. (1.38) is satisfied.

1.4 Ideal thin lens as an optical Fourier transformer

An *ideal thin lens* is a phase object, which means that it will only affect the phase of the incident light. For an ideal converging lens with a focal length f, the phase function of the lens is given by

$$t_f(x,y) = \exp\left[\frac{jk_0}{2f}\left(x^2+y^2\right)\right], \tag{1.41}$$

where we have assumed that the lens is of infinite extent. For a plane wave of amplitude A incident upon the lens, we can employ the Fresnel diffraction formula to calculate the field distribution in the back focal plane of the lens. Using Eq. (1.37) for $z = f$, we have

$$\psi_p(x,y;f) = \exp(-jk_0 f)\frac{jk_0}{2\pi f}\exp\left[\frac{-jk_0}{2f}\left(x^2+y^2\right)\right] \times \mathcal{F}\left\{\psi_{p0}(x,y)\exp\left[\frac{-jk_0}{2f}\left(x^2+y^2\right)\right]\right\}\Bigg|_{k_x=\frac{k_0 x}{f},\,k_y=\frac{k_0 y}{f}}, \tag{1.42}$$

where $\psi_{p0}(x, y)$ is given by $\psi_{p0}(x, y) = A \times t(x, y)$, the amplitude of the incident plane wave multiplied by the transparency function of the aperture. In the present case, the transparency function of the aperture is given by the lens function $t_f(x, y)$, i.e., $t(x, y) = t_f(x, y)$. Hence $\psi_{p0}(x, y) = A \times t(x, y) = A \times t_f(x, y)$. The field distribution f away from the lens, according to Eq. (1.42), is then given by

$$\psi_p(x,y;f) = \exp(-jk_0 f)\frac{jk_0}{2\pi f}\exp\left[\frac{-jk_0}{2f}\left(x^2+y^2\right)\right] \times \mathcal{F}\left\{A\exp\left[\frac{jk_0}{2f}\left(x^2+y^2\right)\right]\exp\left[\frac{-jk_0}{2f}\left(x^2+y^2\right)\right]\right\}\Bigg|_{k_x=\frac{k_0 x}{f},\,k_y=\frac{k_0 y}{f}} \propto \delta(x,y). \tag{1.43}$$

We see that the lens phase function cancels out exactly the quadratic phase function associated with Fresnel diffraction, giving the Fourier transform of constant A proportional to a delta function, $\delta(x, y)$, which is consistent with the geometrical optics which states that all input rays parallel to the optical axis converge behind the lens to a point called the back focal point. The discussion thus far in a sense justifies the functional form of the phase function of the lens given by Eq. (1.41).

We now look at a more complicated situation shown in Fig. 1.6, where a transparency $t(x, y)$ illuminated by a plane wave of unity amplitude is located in the front focal plane of the ideal thin lens.

We want to find the field distribution in the back focal plane. The field immediately after $t(x, y)$ is given by $1 \times t(x, y)$. The resulting field is then

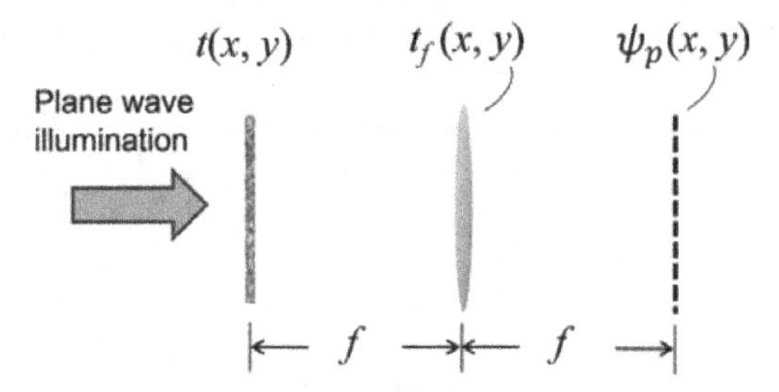

Figure 1.6 Lens as an optical Fourier transformer.

undergoing Fresnel diffraction of a distance f. According to Fresnel diffraction and hence using Eq. (1.35), the diffracted field immediately in front of the lens is given by $t(x, y) * h(x, y; f)$. The field after the lens is then $[t(x, y) * h(x, y; f)] \times t_f(x, y)$. Finally, the field at the back focal plane is found using Fresnel diffraction one more time for a distance of f, as illustrated in Fig. 1.6. The resulting field on the back focal plane of the lens can be written in terms of a series of convolution and multiplication operations as follows [2]:

$$\psi_p(x, y) = \{[t(x, y) * h(x, y; f)]t_f(x, y)\} * h(x, y; f). \tag{1.44}$$

The above equation can be evaluated to become, apart from some constant,

$$\psi_p(x, y) = \mathcal{F}\{t(x, y)\}_{k_x = \frac{k_0 x}{f}, k_y = \frac{k_0 y}{f}} = T\left(\frac{k_0 x}{f}, \frac{k_0 y}{f}\right), \tag{1.45}$$

where $T(k_0x/f, k_0y/f)$ is the Fourier transform or the *spectrum* of $t(x, y)$. We see that we have the exact Fourier transform of the "input," $t(x, y)$, on the back focal plane of the lens. Hence an ideal thin lens is an optical Fourier transformer.

1.5 Optical image processing

Figure 1.6 is the backbone of an optical image processing system. Figure 1.7 shows a standard image processing system with Fig. 1.6 as the front end of the system. The system is known as the 4-f system as lens L_1 and lens L_2 both have the same focal length, f. $p(x, y)$ is called the *pupil function* of the optical system and it is on the confocal plane.

On the object plane, we have an input of the form of a transparency, $t(x, y)$, which is assumed to be illuminated by a plane wave. Hence, according to Eq. (1.45), we have its spectrum on the back focal plane of lens L_1, $T(k_0x/f, k_0y/f)$, where T is the Fourier transform of $t(x, y)$. Hence the confocal plane of the optical system is often called the *Fourier plane*. The spectrum of the input image is now modified by the pupil function as the field immediately after the pupil function

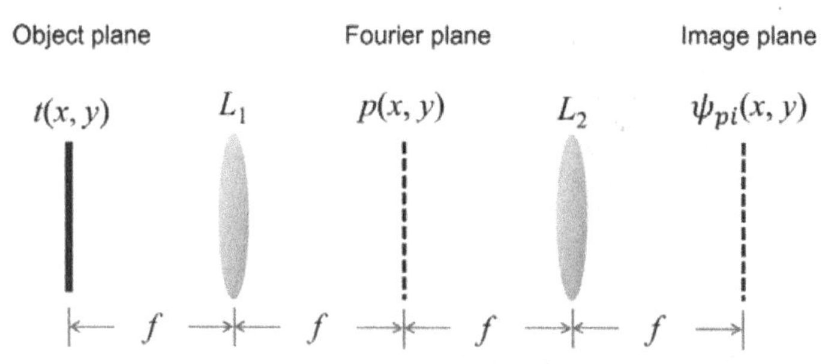

Figure 1.7 4-f image processing system.

is $T(k_0x/f,\ k_0y/f)p(x,\ y)$. According to Eq. (1.45) again, this field will be Fourier transformed to give the field on the image plane as

$$\psi_{pi} = \mathcal{F}\left\{T\left(\frac{k_0x}{f}, \frac{k_0y}{f}\right)p(x,y)\right\}_{k_x=\frac{k_0x}{f},\, k_y=\frac{k_0y}{f}}, \tag{1.46}$$

which can be evaluated, in terms of convolution, to give

$$\begin{aligned}\psi_{pi}(x,y) &= t(-x,-y) * \mathcal{F}\{p(x,\ y)\}_{k_x=\frac{k_0x}{f},\quad k_y=\frac{k_0y}{f}} \\ &= t(-x,-y) * P\left(\frac{k_0x}{f}, \frac{k_0y}{f}\right) \\ &= t(-x,-y) * h_c(x,y),\end{aligned} \tag{1.47}$$

where the negative sign in the argument of $t(x,\ y)$ shows that the original input on the object plane has been flipped and inverted on the image plane. P is the Fourier transform of p. From Eq. (1.47), we define $h_c(x,\ y)$ as the *coherent point spread function (CPSF)* in optics, which is given by [1]

$$h_c(x,y) = \mathcal{F}\{p(x,y)\}_{k_x=\frac{k_0x}{f},\, k_y=\frac{k_0y}{f}} = P\left(\frac{k_0x}{f}, \frac{k_0y}{f}\right). \tag{1.48}$$

By definition, the Fourier transform of the coherent point spread function is the *coherent transfer function (CTF)* given by [1]

$$H_c(k_x,k_y) = \mathcal{F}\{h_c(x,y)\} = \mathcal{F}\left\{P\left(\frac{k_0x}{f}, \frac{k_0y}{f}\right)\right\} = p\left(\frac{-fk_x}{k_0}, \frac{-fk_y}{k_0}\right). \tag{1.49}$$

The expression given by Eq. (1.47) can be interpreted as the flipped and inverted image of $t(x,\ y)$ being processed by the coherent point spread function given by Eq. (1.48). Therefore, image processing capabilities can be varied by simply designing the pupil function, $p(x,\ y)$. Or we can interpret this in the spatial frequency domain as *spatial filtering* is proportional to the functional form of the

Table 1.4 *Properties of a delta function*

Unit area property	$\iint_{-\infty}^{\infty} \delta(x-x_0, y-y_0)dx\,dy = 1$
Scaling property	$\delta(ax, by) = \frac{1}{\lvert ab \rvert}\delta(x, y)$
Product property	$f(x, y)\,\delta(x - x_0, y - y_0) = f(x_0, y_0)\delta(x - x_0, y - y_0)$
Sampling property	$\iint_{-\infty}^{\infty} f(x, y)\delta(x-x_0, y-y_0)dx\,dy = f(x_0, y_0)$

pupil function as evidenced by Eq. (1.46) together with Eq. (1.49). Indeed, Eq. (1.46) is the backbone of so-called *coherent image processing* in optics [1].

Let us look at an example. If we take $p(x, y) = 1$, this means that we do not modify the spectrum of the input image according to Eq. (1.46). Or from Eq. (1.49), the coherent transfer function becomes unity, i.e., all-pass filtering for all spatial frequencies of the input image. Mathematically, using Eq. (1.48) and item number 8 of Table 1.1, $h_c(x, y)$ becomes

$$h_c(x, y) = \mathcal{F}\{1\}_{k_x = \frac{k_0 x}{f}, k_y = \frac{k_0 y}{f}} = 4\pi^2 \delta\left(\frac{k_0 x}{f}, \frac{k_0 y}{f}\right) = 4\pi^2 \left(\frac{f}{k_0}\right)^2 \delta(x, y),$$

a delta function, and the output image from Eq. (1.47) is

$$\psi_{pi}(x, y) \propto t(-x, -y) * \delta\left(\frac{k_0 x}{f}, \frac{k_0 y}{f}\right) \propto t(-x, -y). \tag{1.50}$$

To obtain the last step of the result in Eq. (1.50), we have used the properties of $\delta(x, y)$ in Table 1.4.

If we now take $p(x, y) = \mathrm{circ}(r/r_0)$, from the interpretation of Eq. (1.49) we see that, for this kind of chosen pupil, filtering is of lowpass characteristic as the opening of the circle on the pupil plane only allows the low spatial frequencies to physically go through. Figure 1.8 shows examples of lowpass filtering. In Fig. 1.8 (a) and 1.8(b), we show the original of the image and its spectrum, respectively. In Fig. 1.8(c) and 1.8(e) we show the filtered images, and lowpass filtered spectra are shown in Fig. 1.8(d) and 1.8(f), respectively, where the lowpass filtered spectra are obtained by multiplying the original spectrum by $\mathrm{circ}(r/r_0)$ [see Eq. (1.46)]. Note that the radius r_0 in Fig. 1.8(d) is larger than that in Fig. 1.8(f). In Fig. 1.9, we show highpass filtering examples where we take $p(x, y) = 1 - \mathrm{circ}(r/r_0)$.

So far, we have discussed the use of coherent light, such as plane waves derived from a laser, to illuminate $t(x, y)$ in the optical system shown in Fig. 1.7. The optical system is called a *coherent optical system* in that complex quantities are

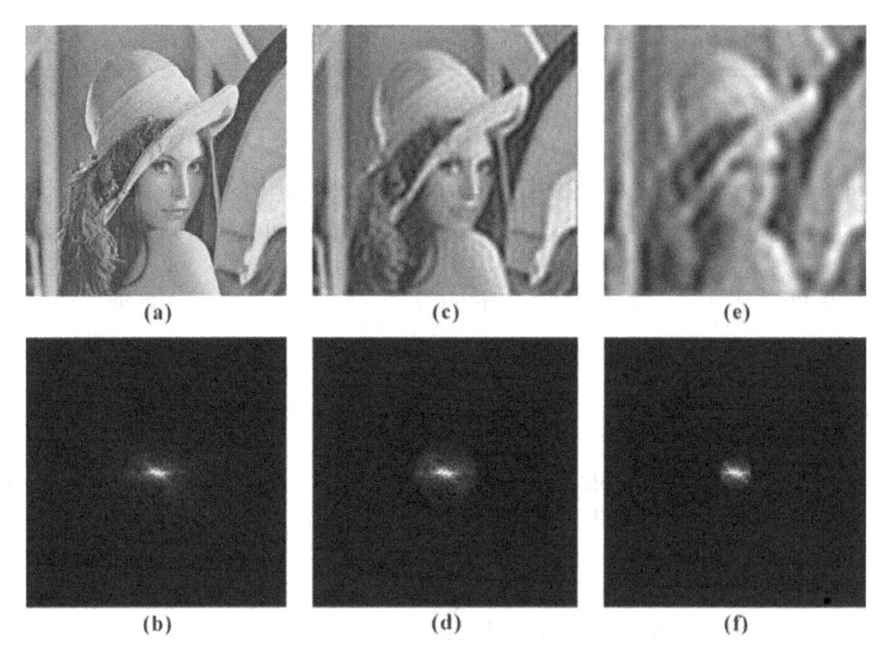

Figure 1.8 Lowpass filtering examples: (a) original image, (b) spectrum of (a); (c) and (e) lowpass images; (d) and (f) spectra of (c) and (e), respectively. See Table 1.5 for the MATLAB code.

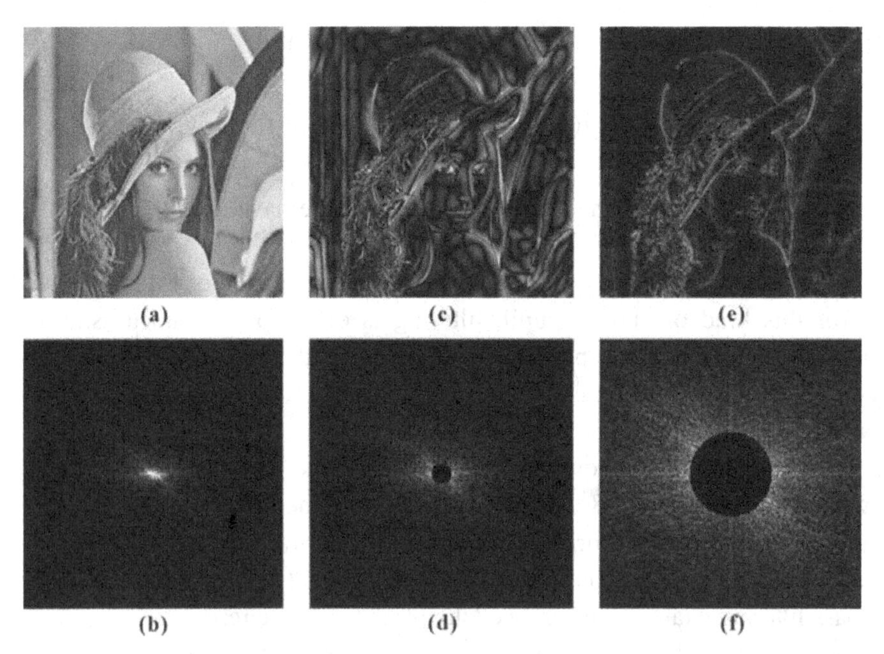

Figure 1.9 Highpass filtering examples: (a) original image, (b) spectrum of (a); (c) and (e) highpass images; (d) and (f) spectra of (c) and (e), respectively. See Table 1.6 for the MATLAB code.

Table 1.5 *MATLAB code for lowpass filtering of an image, see Fig. 1.8*

```
clear all;close all;
A=imread('lena.jpg');    %read 512×512 8bit image
A=double(A);
A=A/255;
SP=fftshift(fft2(fftshift(A)));
D=abs(SP);
D=D(129:384,129:384);
figure;imshow(A);
title('Original image')
figure;imshow(30.*mat2gray(D)); %spectrum
title('Original spectrum')
c=1:512;
r=1:512;
[C, R]=meshgrid(c, r);
CI=((R-257).^2+(C-257).^2);
filter=zeros(512,512);
% produce a high-pass filter
for a=1:512;
   for b=1:512;
      if CI(a,b)>=20^2; %filter diameter
         filter(a,b)=0;
      else
         filter(a,b)=1;
      end
   end
end
G=abs(filter.*SP);
G=G(129:384,129:384);
figure;imshow(30.*mat2gray(G));
title('Low-pass spectrum')
SPF=SP.*filter;
E=abs(fftshift(ifft2(fftshift(SPF))));
figure;imshow(mat2gray(E));
title('Low-pass image')
```

manipulated. Once we have found the complex field on the image plane given by Eq. (1.47), the corresponding image intensity is

$$I_i(x,y) = \psi_{pi}(x,y)\psi_{pi}^*(x,y) = |t(-x,-y) * h_c(x,y)|^2, \tag{1.51}$$

which is the basis for coherent image processing. However, light from extended sources, such as fluorescent tube lights, is incoherent. The system shown in Fig. 1.7 becomes an *incoherent optical system* upon illumination from an incoherent source.

Table 1.6 *MATLAB code for highpass filtering of an image, see Fig. 1.9*

```
clear all;close all;
A=imread('lena.jpg'); %read 512×512 8bit image
A=double(A);
A=A/255;
SP=fftshift(fft2(fftshift(A)));
D=abs(SP);
D=D(129:384,129:384);
figure;imshow(A);
title('Original image')
figure;imshow(30.*mat2gray(D)); % spectrum
title('Original spectrum')
c=1:512;
r=1:512;
[C, R]=meshgrid(c, r);
CI=((R-257).^2+(C-257).^2);
filter=zeros(512,512);
% produce a high-pass filter
for a=1:512;
   for b=1:512;
      if CI(a,b)<=20^2; %filter diameter
         filter(a,b)=0;
      else
         filter(a,b)=1;
      end
   end
end
G=abs(filter.*SP);
G=G(129:384,129:384);
figure;imshow(2.*mat2gray(G));
title('High-pass spectrum')
SPF=SP.*filter;
E=abs(fftshift(ifft2(fftshift(SPF))));
figure;imshow(mat2gray(E));
title('High-pass image')
```

The optical system manipulates intensity quantities directly. To find the image intensity, we perform convolution with the given intensity quantities as follows:

$$I_i(x,y) = |t(-x,-y)|^2 * |h_c(x,y)|^2. \tag{1.52}$$

Equation (1.52) is the basis for *incoherent image processing* [1], and $|h_c(x, y)|^2$ is the *intensity point spread function (IPSF)* [1]. Note that the IPSF is real and non-negative, which means that it is not possible to implement even the simplest enhancement and restoration algorithms (e.g., highpass, derivatives, etc.), which

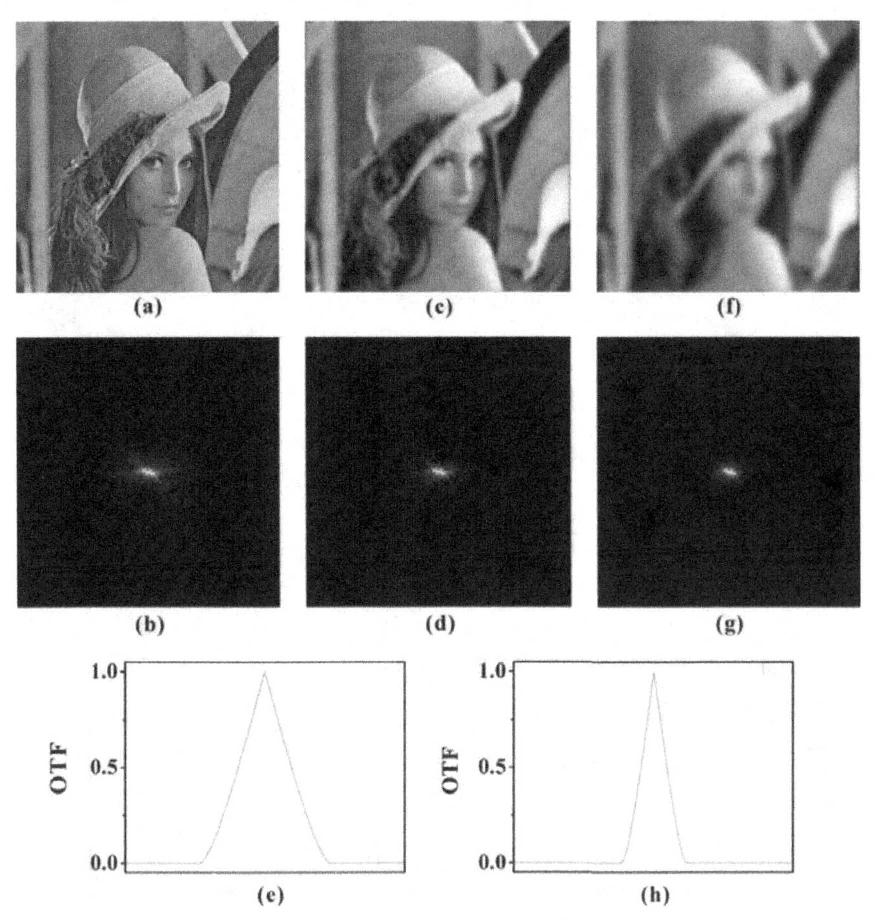

Figure 1.10 Incoherent spatial filtering examples using $p(x, y) = \text{circ}(r / r_0)$: (a) original image, (b) spectrum of (a); (c) and (f) filtered images; (d) and (g) spectra of (c) and (f), respectively; (e) and (h) cross sections through the center of the OTF using different r_0 in the pupil function for the processed images in (c) and (f), respectively. The full dimension along the horizontal axis contains 256 pixels for figures (b), (d) and (g), while figures (e) and (h) zoom in the peak with 30 pixels plotted. See Table 1.7 for the MATLAB code.

require a *bipolar point spread function.* Novel incoherent image processing techniques seek to realize bipolar point spread functions (see, for example, [3–6]).

The Fourier transform of the IPSF gives a transfer function known as the *optical transfer function (OTF)* of the incoherent optical system:

$$\text{OTF}(k_x, k_y) = \mathcal{F}\left\{|h_c(x, y)|^2\right\}. \tag{1.53}$$

Using Eq. (1.49), we can relate the coherent transfer function to the OTF as follows:

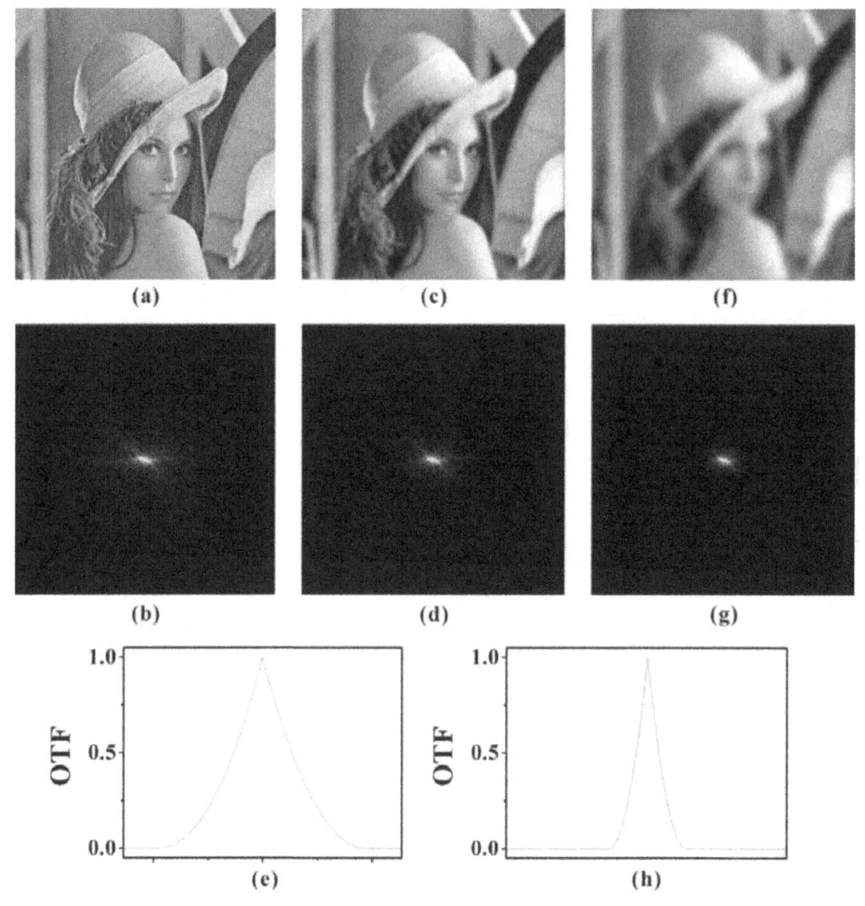

Figure 1.11 Incoherent spatial filtering examples using $p(x, y) = 1 - \text{circ}(r / r_0)$: (a) original image, (b) spectrum of (a); (c) and (f) filtered images; (d) and (g) spectra of (c) and (f), respectively; (e) and (h) cross sections through the center of the OTF using different r_0 in the pupil function for the processed images of (c) and (f), respectively. The full dimension along x contains 256 pixels for figures (b), (d) and (g), while figures (e) and (h) zoom in the peak with 30 pixels plotted. See Table 1.8 for the MATLAB code.

$$\begin{aligned}\text{OTF}(k_x, k_y) &= H_c(k_x, k_y) \otimes H_c(k_x, k_y) \\ &= \iint_{-\infty}^{\infty} H_c^*(k_x', k_y') H_c(k_x' + k_x, k_y' + k_y) dk_x' dk_y',\end{aligned} \tag{1.54}$$

where $\otimes$ defines correlation [see Table 1.1]. The modulus of the OTF is called the *modulation transfer function* (*MTF*), and it is important to note that

$$|\text{OTF}(k_x, k_y)| \leq |\text{OTF}(0, 0)|, \tag{1.55}$$

Table 1.7 *MATLAB code for incoherent spatial filtering, circ(r/r₀), see Fig. 1.10*

```
clear all;close all;
A=imread('lena.jpg'); %read 512×512 8bit image
A=double(A);
A=A/255;
SP=fftshift(fft2(fftshift(A)));
D=abs(SP);
D=D(129:384,129:384);
figure;imshow(A);
title('Original image')
figure;imshow(30.*mat2gray(D)); %spectrum
title('Original spectrum')
c=1:512;
r=1:512;
[C, R]=meshgrid(c, r);
CI=((R-257).^2+(C-257).^2);
pup=zeros(512,512);
% produce a circular pupil
for a=1:512;
    for b=1:512;
        if CI(a,b)>=30^2; %pupil diameter 30(15)
            pup(a,b)=0;
        else
            pup(a,b)=1;
        end
    end
end
h=ifft2(fftshift(pup));
OTF=fftshift(fft2(h.*conj(h)));
OTF=OTF/max(max(abs(OTF)));
G=abs(OTF.*SP);
G=G(129:384,129:384);
figure;imshow(30.*mat2gray(G));
title('Filtered spectrum')
I=abs(fftshift(ifft2(fftshift(OTF.*SP))));
figure;imshow(mat2gray(I));
title('Filtered image')
```

which states that the MTF always has a central maximum. This signifies that we always have lowpass filtering characteristics regardless of the pupil function used in an incoherent optical system. In Figs. 1.10 and 1.11, we show *incoherent spatial filtering* results in an incoherent optical system [1] using $p(x, y) = \mathrm{circ}(r/r_0)$ and $p(x, y) = 1 - \mathrm{circ}(r/r_0)$, respectively.

Table 1.8 *MATLAB code for incoherent spatial filtering, 1−circ(r/r₀), see Fig. 1.11*

```
clear all;close all;
A=imread('lena.jpg'); %512×512 8bit image
A=double(A);
A=A/255;
SP=fftshift(fft2(fftshift(A)));
D=abs(SP);
D=D(129:384,129:384);
figure;imshow(A);
title('Original image')
figure;imshow(30.*mat2gray(D)); %spectrum
title('Original spectrum')
c=1:512;
r=1:512;
[C, R]=meshgrid(c, r);
CI=((R-257).^2+(C-257).^2);
pup=zeros(512,512);
% produce a circular pupil
for a=1:512;
   for b=1:512;
      if CI(a,b)>=350^2; %pupil diameter
         pup(a,b)=1;
      else
         pup(a,b)=0;
      end
   end
end
h=ifft2(fftshift(pup));
OTF=fftshift(fft2(h.*conj(h)));
OTF=OTF/max(max(abs(OTF)));
G=abs(OTF.*SP);
G=G(129:384,129:384);
figure;imshow(30.*mat2gray(G));
title('Filtered spectrum')
I=abs(fftshift(ifft2(fftshift(OTF.*SP))));
figure;imshow(mat2gray(I));
title('Filtered image')
```

Problems

1.1 Starting from the Maxwell's equations, (a) derive the wave equation for $\boldsymbol{E}$ in a linear, homogeneous, and isotropic medium characterized by ε and μ, and (b) do the same as in (a) but for $\boldsymbol{H}$.

1.2 Verify the Fourier transform properties 2, 3 and 4 in Table 1.1.

1.3 Verify the Fourier transform pairs 5 and 6 in Table 1.1.

1.4 Verify the Fourier transform pairs 7, 8, 9, 10, and 11 in Table 1.1.

1.5 Assume that the solution to the three-dimensional wave equation in Eq. (1.11) is given by $\psi(x, y, z, t) = \psi_p(x, y; z)\exp(j\omega_0 t)$, verify that the Helmholtz equation for $\psi_p(x, y; z)$ is given by

$$\frac{\partial^2 \psi_p}{\partial x^2} + \frac{\partial^2 \psi_p}{\partial y^2} + \frac{\partial^2 \psi_p}{\partial z^2} + k_0^2 \psi_p = 0,$$

where $k_0 = \omega_0 / v$.

1.6 Write down functions of the following physical quantities in Cartesian coordinates (x, y, z).
(a) A plane wave on the x–z plane in free space. The angle between the propagation vector and the z-axis is θ.
(b) A diverging spherical wave emitted from a point source at (x_0, y_0, z_0) under paraxial approximation.

1.7 A rectangular aperture described by the transparency function $t(x, y) = \mathrm{rect}(x/x_0, y/y_0)$ is illuminated by a plane wave of unit amplitude. Determine the complex field, $\psi_p(x, y; z)$, under Fraunhofer diffraction. Plot the intensity, $|\psi_p(x, y; z)|^2$, along the x-axis and label all essential points along the axis.

1.8 Repeat Problem 1.7 but with the transparency function given by

$$t(x,y) = \mathrm{rect}\left(\frac{x - X/2}{x_0}, \frac{y}{y_0}\right) + \mathrm{rect}\left(\frac{x + X/2}{x_0}, \frac{y}{y_0}\right), \qquad X \gg x_0.$$

1.9 Assume that the pupil function in the 4-f image processing system in Fig. 1.7 is given by $\mathrm{rect}(x/x_0, y/y_0)$. (a) Find the coherent transfer function, (b) give an expression for the optical transfer function and express it in terms of the coherent transfer function, and (c) plot both of the transfer functions.

1.10 Repeat Problem 1.9 but with the pupil function given by the transparency in Problem 1.8.

1.11 Consider a grating with transparency function $t(x, y) = \frac{1}{2} + \frac{1}{2}\cos(2\pi x/\Lambda)$, where Λ is the period of the grating. Determine the complex field, $\psi_p(x, y; z)$, under Fresnel diffraction if the grating is normally illuminated by a unit amplitude plane wave.

1.12 Consider the grating given in Problem 1.11. Determine the complex field, $\psi_p(x, y; z)$, under Fraunhofer diffraction if the grating is normally illuminated by a unit amplitude plane wave.

1.13 Consider the grating given in Problem 1.11 as the input pattern in the 4-f image processing system in Fig. 1.7. Assuming coherent illumination, find the intensity distribution at the output plane when a small opaque stop is located at the center of the Fourier plane.

References

1. T.-C. Poon, and P. P. Banerjee, *Contemporary Optical Image Processing with MATLAB®* (Elsevier, Oxford, UK, 2001).
2. T.-C. Poon, and T. Kim, *Engineering Optics with MATLAB®* (World Scientific, River Hackensack, NJ, 2006).
3. A. W. Lohmann, and W. T. Rhodes, Two-pupil synthesis of optical transfer functions, *Applied Optics* **17**, 1141–1151 (1978).
4. W. Stoner, Incoherent optical processing via spatially offset pupil masks, *Applied Optics* **17**, 2454–2467 (1978).
5. T.-C. Poon, and A. Korpel, Optical transfer function of an acousto-optic heterodyning image processor, *Optics Letters* **4**, 317–319 (1979).
6. G. Indebetouw, and T.-C. Poon, Novel approaches of incoherent image processing with emphasis on scanning methods, *Optical Engineering* **31**, 2159–2167 (1992).

2

Fundamentals of holography

2.1 Photography and holography

When an object is illuminated, we see the object as light is scattered to create an *object wave* reaching our eyes. The object wave is characterized by two quantities: the *amplitude*, which corresponds to brightness or intensity, and the *phase*, which corresponds to the shape of the object. The amplitude and phase are conveniently represented by the so-called complex amplitude introduced in Chapter 1. The complex amplitude contains complete information about the object. When the object wave illuminates a recording medium such as a photographic film or a CCD camera, what is recorded is the variation in light intensity at the plane of the recording medium as these recording media respond only to light intensity. Mathematically, the intensity, $I(x, y)$, is proportional to the complex amplitude squared, i.e., $I(x, y) \propto |\psi_p(x, y)|^2$, where ψ_p is the two-dimensional complex amplitude on the recording medium. The result of the variation in light intensity is a *photograph* and if we want to make a transparency from it, the amplitude transmittance $t(x, y)$ of the transparency can be made proportional to the recorded intensity, or we simply write as follows:

$$t(x, y) = |\psi_p(x, y)|^2. \tag{2.1}$$

Hence in photography, as a result of this intensity recording, all information about the relative phases of the light waves from the original three-dimensional scene is lost. This loss of the phase information of the light field in fact destroys the three-dimensional character of the scene, i.e., we cannot change the perspective of the image in the photograph by viewing it from a different angle (i.e., parallax) and we cannot interpret the depth of the original three-dimensional scene. In essence, a photograph is a two-dimensional recording of a three-dimensional scene.

Holography is a method invented by Gabor in 1948 [1] in which not only the amplitude but also the phase of the light field can be recorded. The word "holography" combines parts of two Greek words: holos, meaning "complete," and graphein, meaning "to write" or "to record." Thus, holography means the recording of complete information. Hence, in the holographic process, the recording medium records the original complex amplitude, i.e., both the amplitude and phase of the complex amplitude of the object wave. The result of the recorded intensity variations is now called a *hologram*. When a hologram is properly illuminated at a later time, our eyes observe the intensity generated by the same complex field. As long as the exact complex field is restored, we can observe the original complex field at a later time. The restored complex field preserves the entire parallax and depth information much like the original complex field and is interpreted by our brain as the same three-dimensional object.

2.2 Hologram as a collection of Fresnel zone plates

The principle of holography can be explained by recording a point object since any object can be considered as a collection of points. Figure 2.1 shows a collimated laser split into two plane waves and recombined through the use of two mirrors (M1 and M2) and two beam splitters (BS1 and BS2).

One plane wave is used to illuminate the pinhole aperture (our point object), and the other illuminates the recording medium directly. The plane wave that is

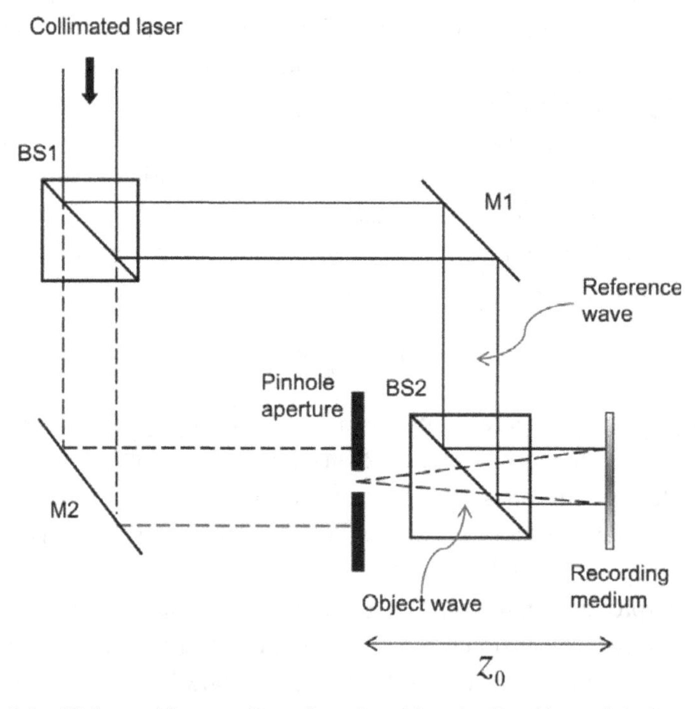

Figure 2.1 Holographic recording of a point object (realized by a pinhole aperture).

scattered by the point object generates a diverging spherical wave toward the recording medium. This diverging wave is known as an *object wave* in holography. The plane wave that directly illuminates the recording medium is known as a *reference wave*. Let ψ_0 represent the field distribution of the object wave on the plane of the recording medium, and similarly let ψ_r represent the field distribution of the reference wave on the plane of the recording medium. The recording medium now records the interference of the reference wave and the object wave, i.e., what is recorded is given by $|\psi_0 + \psi_r|^2$, provided the reference wave and the object wave are mutually coherent over the recording medium. The coherence of the light waves is guaranteed by the use of a laser source (we will discuss coherence in Section 2.4). This kind of recording is known as holographic recording, distinct from a photographic recording in that the reference wave does not exist and hence only the object wave is recorded.

We shall discuss holographic recording of a point source object mathematically. Let us consider the recording of a point object at a distance z_0 from the recording medium as shown in Fig. 2.1. The pinhole aperture is modeled as a delta function, $\delta(x, y)$, which gives rise to an object wave, ψ_0, according to Fresnel diffraction [see Eq. (1.35)], on the recording medium as

$$\begin{aligned}\psi_0(x, y; z_0) &= \delta(x, y) * h(x, y; z_0) = \delta(x, y) * \exp(-jk_0 z_0)\frac{jk_0}{2\pi z_0}\exp\left[\frac{-jk_0}{2z_0}(x^2 + y^2)\right] \\ &= \exp(-jk_0 z_0)\frac{jk_0}{2\pi z_0}\exp\left[\frac{-jk_0}{2z_0}(x^2 + y^2)\right]. \qquad (2.2)\end{aligned}$$

This object wave is a *paraxial spherical wave*. For the reference plane wave, we assume that the plane wave has the same initial phase as the point object at a distance z_0 away from the recording medium. Therefore, its field distribution on the recording medium is $\psi_r = a \exp(-jk_0 z_0)$, where a, considered real for simplicity here, is the amplitude of the plane wave. Hence, the recorded intensity distribution on the recording medium or the hologram with amplitude transmittance is given by

$$t(x, y) = |\psi_r + \psi_0|^2 = \left| a \exp(-jk_0 z_0) + \exp(-jk_0 z_0)\frac{jk_0}{2\pi z_0}\exp\left[\frac{-jk_0}{2z_0}(x^2 + y^2)\right]\right|^2$$

or

$$t(x, y) = a^2 + \left(\frac{k_0}{2\pi z_0}\right)^2 + a\frac{-jk_0}{2\pi z_0}\exp\left[\frac{jk_0}{2z_0}(x^2 + y^2)\right] + a\frac{jk_0}{2\pi z_0}\exp\left[\frac{-jk_0}{2z_0}(x^2 + y^2)\right]. \qquad (2.3)$$

Note that the last term, which is really the desirable term of the equation, is the total complex field of the original object wave [see Eq. (2.2)] aside from the constant

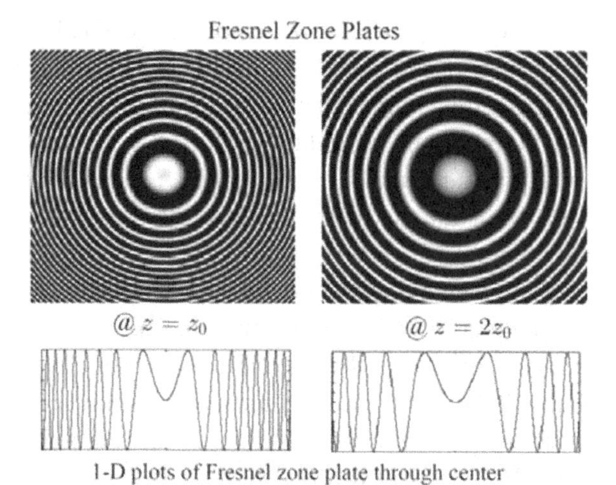

Figure 2.2 Plots of Fresnel zone plates for $z = z_0$ and $z = 2z_0$.

terms a and $\exp(-jk_0z_0)$. Now, Eq. (2.3) can be simplified to a real function and we have a real hologram given by

$$t(x, y) = A + B \sin\left[\frac{k_0}{2z_0}(x^2 + y^2)\right], \tag{2.4}$$

where $A = a^2 + (k_0/2\pi z_0)^2$ is some constant bias, and $B = ak_0/\pi z_0$. The expression in Eq. (2.4) is often called the sinusoidal *Fresnel zone plate (FZP)*, which is the hologram of the point source object at a distance $z = z_0$ away from the recording medium. Plots of the FZPs are shown in Fig. 2.2, where we have set k_0 to be some constant but for $z = z_0$ and $z = 2z_0$.

When we investigate the quadratic spatial dependence of the FZP, we notice that the spatial rate of change of the phase of the FZP, say along the x-direction, is given by

$$f_{local} = \frac{1}{2\pi}\frac{d}{dx}\left(\frac{k_0}{2z_0}x^2\right) = \frac{k_0 x}{2\pi z_0}. \tag{2.5}$$

This is a *local fringe frequency* that increases linearly with the spatial coordinate, x. In other words, the further we are away from the center of the zone, the higher the local spatial frequency, which is obvious from Fig. 2.2. Note also from the figure, when we double the z value, say from $z = z_0$ to $z = 2z_0$, the local fringes become less dense as evident from Eq. (2.5) as well. Hence the local frequency carries the information on z, i.e., from the local frequency we can deduce how far the object point source is away from the recording medium – an important aspect of holography.

To reconstruct the original light field from the hologram, $t(x, y)$, we can simply illuminate the hologram with plane wave ψ_{rec}, called the *reconstruction wave* in holography, which gives a complex amplitude at z away from the hologram, according to Fresnel diffraction,

$$\psi_{rec}t(x,y) * h(x,y;z) = \psi_{rec}\left\{A + a\frac{-jk_0}{2\pi z_0}\exp\left[\frac{jk_0}{2z_0}(x^2+y^2)\right]\right.$$
$$\left. +a\frac{jk_0}{2\pi z_0}\exp\left[\frac{-jk_0}{2z_0}(x^2+y^2)\right]\right\} * h(x,y;z). \quad (2.6)$$

Evaluation of the above equation gives three light fields emerging from the hologram. The light field due to the first term in the hologram is a plane wave as $\psi_{rec}A * h(x, y; z) \propto \psi_{rec}A$, which makes sense as the plane wave propagates without diffraction. This out-going plane wave is called a *zeroth-order beam* in holography, which provides a uniform output at the observation plane. In the present analysis the interference is formed using a paraxial spherical wave and a plane wave. So the zeroth-order beam is uniform. However, if the object light has some amplitude variations, the zeroth-order beam will not be uniform. Now, the field due to the second term is

$$\psi_{rec}a\frac{-jk_0}{2\pi z_0}\exp\left[\frac{jk_0}{2z_0}(x^2+y^2)\right] * h(x,y;z)$$
$$\propto \frac{-jk_0}{2\pi z_0}\frac{jk_0}{2\pi z}\exp\left[\frac{jk_0}{2z_0}(x^2+y^2)\right] * \exp\left[\frac{-jk_0}{2z}(x^2+y^2)\right]$$
$$= \frac{-jk_0}{2\pi z_0}\frac{jk_0}{2\pi z}\exp\left[\frac{jk_0}{2(z_0-z)}(x^2+y^2)\right]. \quad (2.7)$$

This is a converging spherical wave if $z < z_0$. However, when $z > z_0$, we have a diverging wave. For $z = z_0$, the wave focuses to a real point source z_0 away from the hologram and is given by a delta function, $\delta(x, y)$. Now, finally for the last term in the equation, we have

$$\psi_{rec}a\frac{jk_0}{2\pi z_0}\exp\left[\frac{-jk_0}{2z_0}(x^2+y^2)\right] * h(x,y;z)$$
$$\propto \frac{jk_0}{2\pi z_0}\frac{jk_0}{2\pi z}\exp\left[\frac{-jk_0}{2z_0}(x^2+y^2)\right] * \exp\left[\frac{-jk_0}{2z}(x^2+y^2)\right]$$
$$= \frac{jk_0}{2\pi z_0}\frac{jk_0}{2\pi z}\exp\left[\frac{-jk_0}{2(z_0+z)}(x^2+y^2)\right], \quad (2.8)$$

and we have a diverging wave with its virtual point source at a distance $z = -z_0$, which is behind the hologram, on the opposite side to the observer. This reconstructed point source is at the exact location of the original point source object.

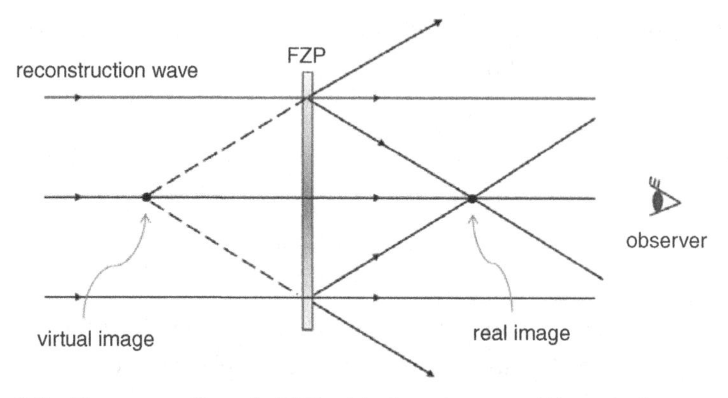

Figure 2.3 Reconstruction of a FZP with the existence of the twin image (which is the real image reconstructed in the figure).

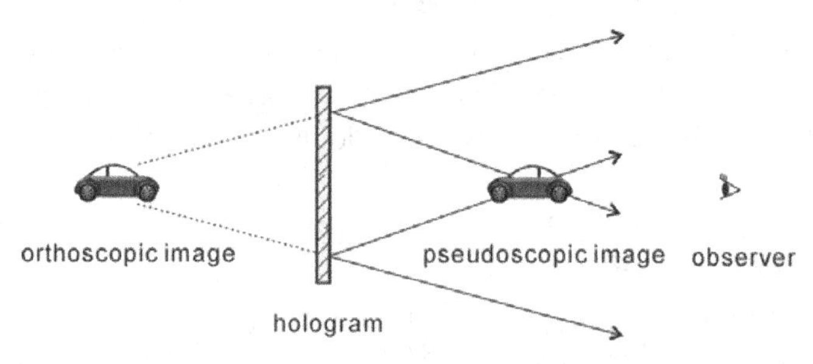

Figure 2.4 Orthoscopic and pseudoscopic images.

The situation is illustrated in Fig. 2.3. The reconstructed real point source is called the *twin image* of the virtual point source.

Although both the virtual image and the real image exhibit the depth of the object, the virtual image is usually used for applications of three-dimensional display. For the virtual image, the observer will see a reconstructed image with the same perspective as the original object. For the real image, the reconstructed image is a mirror and inside-out image of the original object, as shown in Fig. 2.4. This type of image is called the *pseudoscopic image*, while the image with normal perspective is called the *orthoscopic image*. Because the pseudoscopic image cannot provide natural parallax to the observer, it is not suitable for three-dimensional display.

Let us now see what happens if we have two point source objects given by $\delta(x, y) + \delta(x - x_1, y - y_1)$. They are located z_0 away from the recording medium. The object wave now becomes

$$\psi_0(x, y; z_0) = [b_0\delta(x, y) + b_1\delta(x - x_1, y - y_1)] * h(x, y; z_0), \tag{2.9}$$

where b_0 and b_1 denote the amplitudes of the two point sources. The hologram now becomes

$$t(x,y) = |\psi_r + \psi_0(x,y;z_0)|^2$$
$$= \left| a\exp(-jk_0z_0) + b_0\exp(-jk_0z_0)\frac{jk_0}{2\pi z_0}\exp\left[\frac{-jk_0}{2z_0}(x^2+y^2)\right]\right.$$
$$\left. + b_1\exp(-jk_0z_0)\frac{jk_0}{2\pi z_0}\exp\left\{\frac{-jk_0}{2z_0}\left[(x-x_1)^2+(y-y_1)^2\right]\right\}\right|^2. \quad (2.10)$$

Again, the above expression can be put in a real form, i.e., we have

$$t(x,y) = C + \frac{ab_0k_0}{\pi z_0}\sin\left[\frac{k_0}{2z_0}(x^2+y^2)\right] + \frac{ab_1k_0}{\pi z_0}\sin\left\{\frac{k_0}{2z_0}[(x-x_1)^2+(y-y_1)^2]\right\}$$
$$+ 2b_0b_1\left(\frac{k_0}{2\pi z_0}\right)^2\cos\left\{\frac{k_0}{2z_0}[(x_1^2+y_1^2)+2xx_1+2yy_1]\right\}, \quad (2.11)$$

where C is again some constant bias obtained similarly as in Eq. (2.4). We recognize that the second and third terms are our familiar FZP associated to each point source, while the last term is a cosinusoidal fringe grating which comes about due to interference among the spherical waves. Again, only one term from each of the sinusoidal FZPs contains the desirable information as each contains the original light field for the two points. The other terms in the FZPs are undesirable upon reconstruction, and give rise to twin images. The cosinusoidal grating in general introduces noise on the reconstruction plane. If we assume the two point objects are close together, then the spherical waves reaching the recording medium will intersect at small angles, giving rise to interference fringes of low spatial frequency. This low frequency as it appears on the recording medium corresponds to a coarse grating, which diffracts the light by a small angle, giving the zeroth-order beam some structure physically. Parker Givens has previously given a general form of such a hologram due to a number of point sources [2, 3].

2.3 Three-dimensional holographic imaging

In this section, we study the lateral and longitudinal magnifications in holographic imaging. To make the situation a bit more general, instead of using plane waves for recording and reconstruction as in the previous section, we use point sources. We consider the geometry for recording shown in Fig. 2.5. The two point objects, labeled 1 and 2, and the reference wave, labeled R, emit spherical waves that, on the plane of the recording medium, contribute to complex fields, ψ_{p1}, ψ_{p2}, and ψ_{pR}, respectively, given by

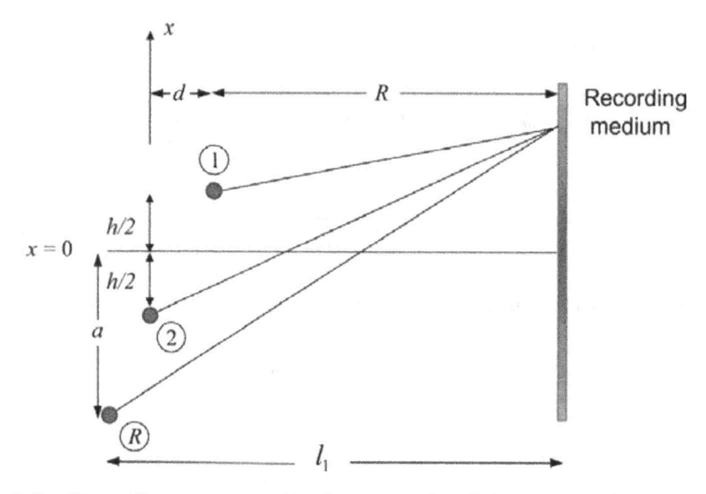

Figure 2.5 Recording geometry for the two point objects, 1 and 2. The reference point source is labeled R.

$$\psi_{p1}(x,y) = \delta\left(x-\frac{h}{2},y\right) * h(x,y;R) = \exp(-jk_0R)\frac{jk_0}{2\pi R}\exp\left\{\frac{-jk_0}{2R}\left[(x-h/2)^2+y^2\right]\right\}$$
$$\propto \exp\left\{\frac{-jk_0}{2R}\left[(x-h/2)^2+y^2\right]\right\}, \tag{2.12}$$

$$\psi_{p2}(x,y) = \delta\left(x+\frac{h}{2},y\right) * h(x,y;R+d)$$
$$= \exp\left[(-jk_0(R+d))\right]\frac{jk_0}{2\pi(R+d)}\exp\left\{\frac{-jk_0}{2(R+d)}\left[(x+h/2)^2+y^2\right]\right\}$$
$$\propto \exp\left\{\frac{-jk_0}{2(R+d)}\left[(x+h/2)^2+y^2\right]\right\}, \tag{2.13}$$

and

$$\psi_{pR}(x,y) = \delta(x+a,y) * h(x,y;l_1) = \exp(-jk_0l_1)\frac{jk_0}{2\pi l_1}\exp\left\{\frac{-jk_0}{2l_1}\left[(x+a)^2+y^2\right]\right\}$$
$$\propto \exp\left\{\frac{-jk_0}{2l_1}\left[(x+a)^2+y^2\right]\right\}. \tag{2.14}$$

These spherical waves interfere on the recording medium to yield a hologram given by

$$\begin{aligned} t(x,y) &= |\psi_{p1}(x,y) + \psi_{p2}(x,y) + \psi_{pR}(x,y)|^2 \\ &= [\psi_{p1}(x,y) + \psi_{p2}(x,y) + \psi_{pR}(x,y)][\psi^*_{p1}(x,y) + \psi^*_{p2}(x,y) + \psi^*_{pR}(x,y)], \end{aligned} \tag{2.15}$$

where the superscript * represents the operation of a complex conjugate. Rather than write down the complete expression for $t(x, y)$ explicitly, we will, on the basis of our previous experience, pick out some relevant terms responsible for image reconstruction. The terms of relevance are $t_{reli}(x, y)$, where $i = 1,2,3,4$

$$t_{rel1}(x,y) = \psi^*_{p1}(x,y)\psi_{pR}(x,y)$$
$$= \exp\left\{\frac{jk_0}{2R}[(x-h/2)^2+y^2]\right\} \times \exp\left\{\frac{-jk_0}{2l_1}[(x+a)^2+y^2]\right\}, \quad (2.16a)$$

$$t_{rel2}(x,y) = \psi^*_{p2}(x,y)\psi_{pR}(x,y)$$
$$= \exp\left\{\frac{jk_0}{2(R+d)}[(x+h/2)^2+y^2]\right\} \times \exp\left\{\frac{-jk_0}{2l_1}[(x+a)^2+y^2]\right\}, \quad (2.16b)$$

$$t_{rel3}(x,y) = \psi_{p1}(x,y)\psi^*_{pR}(x,y) = [t_{rel1}(x,y)]^*, \quad (2.16c)$$

$$t_{rel4}(x,y) = \psi_{p2}(x,y)\psi^*_{pR}(x,y) = [t_{rel2}(x,y)]^*. \quad (2.16d)$$

Note that $t_{rel3}(x, y)$ and $t_{rel4}(x, y)$ contain the original wavefronts $\psi_{p1}(x, y)$ and $\psi_{p2}(x, y)$ of points 1 and 2, respectively, and upon reconstruction they give rise to virtual images as shown in the last section for a single point object. However, $t_{rel1}(x, y)$ and $t_{rel2}(x, y)$ contain the complex conjugates of the original complex amplitudes $\psi^*_{p1}(x,y)$ and $\psi^*_{p2}(x,y)$ of points 1 and 2, respectively, and upon reconstruction they give rise to real images. We shall now show how these reconstructions come about mathematically for spherical reference recording and reconstruction.

The reconstruction geometry for real images is shown in Fig. 2.6, where the hologram just constructed is illuminated with a reconstruction spherical wave from

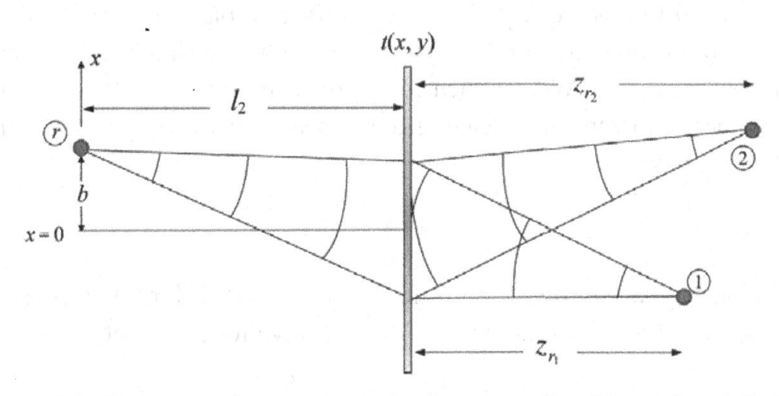

Figure 2.6 Reconstruction geometry for the two point objects, 1 and 2. The reconstruction point source is labeled *r*.

a point source labeled r. For simplicity, we assume that the wavelength of the reconstruction wave is the same as that of the waves of the recording process.

Hence the complex field, $\psi_{pr}(x, y)$, illuminating the hologram is

$$\psi_{pr}(x,y) = \delta(x-b,y) * h(x,y;l_2) = \exp(-jk_0 l_2)\frac{jk_0}{2\pi l_2}\exp\left\{\frac{-jk_0}{2l_2}\left[(x-b)^2+y^2\right]\right\}$$
$$\propto \exp\left\{\frac{-jk_0}{2l_2}\left[(x-b)^2+y^2\right]\right\}. \tag{2.17}$$

We find the total complex field immediately behind (away from the source) the hologram by multiplying Eq. (2.17) with Eq. (2.15) but the reconstructions due to the relevant terms are

$$\psi_{pr}(x,y)t_{reli}(x,y), \tag{2.18}$$

where the t_{reli} are defined in Eqs. (2.16).

Consider, first, the contribution from $\psi_{pr}(x,y)t_{rel1}(x,y)$. After propagation through a distance z behind the hologram, the complex field is transformed according to the Fresnel diffraction formula. Note that because the field is converging, it will contribute to a real image. Explicitly, the field can be written as

$$\psi_{pr}(x,y)t_{rel1}(x,y) * h(x,y;z)$$
$$= \psi_{pr}(x,y)t_{rel1}(x,y) * \exp(-jk_0 z)\frac{jk_0}{2\pi z}\exp\left[\frac{-jk_0}{2z}(x^2+y^2)\right]$$
$$\propto \exp\left\{\frac{-jk_0}{2l_2}[(x-b)^2+y^2]\right\}\exp\left\{\frac{jk_0}{2R}[(x-h/2)^2+y^2]\right\}$$
$$\exp\left\{\frac{-jk_0}{2l_1}[(x+a)^2+y^2]\right\} * \frac{jk_0}{2\pi z}\exp\left[\frac{-jk_0}{2z}(x^2+y^2)\right]. \tag{2.19}$$

From the definition of convolution integral [see Table 1.1], we perform the integration by writing the functions involved with new independent variables x', y' and $(x-x', y-y')$. We can then equate the coefficients of x'^2, y'^2, appearing in the exponents, to zero, thus leaving only linear terms in x', y'. Doing this for Eq. (2.19), we have

$$\frac{1}{R}-\frac{1}{l_1}-\frac{1}{l_2}-\frac{1}{z_{r1}}=0, \tag{2.20}$$

where we have replaced z by z_{r1}. z_{r1} is the distance of the real image reconstruction of point object 1 behind the hologram. We can solve for z_{r1} to get

$$z_{r1} = \left[\frac{1}{R}-\frac{1}{l_1}-\frac{1}{l_2}\right]^{-1} = \frac{Rl_1l_2}{l_1l_2-(l_1+l_2)R}. \tag{2.21}$$

At this distance, we can write Eq. (2.19) as

$$\psi_{pr}(x,y)t_{rel1}(x,y) * h(x,y;z_{r1})$$
$$\propto \iint_{-\infty}^{\infty} \exp\left[jk_0\left(-\frac{h}{2R}-\frac{a}{l_1}+\frac{b}{l_2}+\frac{x}{z_{r1}}\right)x' + jk_0\frac{y}{z_{r1}}y'\right]dx'dy'$$
$$\propto \delta\left[x + z_{r1}\left(\frac{b}{l_2}-\frac{h}{2R}-\frac{a}{l_1}\right), y\right], \qquad (2.22)$$

which is a δ function shifted in the lateral direction and is a real image of point object 1. The image is located z_{r1} away from the hologram and at

$$x = x_1 = -z_{r1}\left(\frac{b}{l_2}-\frac{h}{2R}-\frac{a}{l_1}\right), \qquad y = y_1 = 0.$$

As for the reconstruction due to the relevant term $\psi_{pr}(x,y)t_{rel2}(x,y)$ in the hologram, we have

$$\psi_{pr}(x,y)t_{rel2}(x,y) * h(x,y;z)$$
$$= \psi_{pr}(x,y)t_{rel2}(x,y) * \exp(-jk_0z)\frac{jk_0}{2\pi z}\exp\left[\frac{-jk_0}{2z}(x^2+y^2)\right]$$
$$\propto \exp\left\{\frac{-jk_0}{2l_2}\left[(x-b)^2+y^2\right]\right\}\exp\left\{\frac{jk_0}{2(R+d)}\left[(x+h/2)^2+y^2\right]\right\}$$
$$\exp\left\{\frac{-jk_0}{2l_1}\left[(x+a)^2+y^2\right]\right\} * \frac{jk_0}{2\pi z}\exp\left[\frac{-jk_0}{2z}(x^2+y^2)\right]. \qquad (2.23)$$

A similar analysis reveals that this is also responsible for a real image reconstruction but for point object 2, expressible as

$$\psi_{pr}(x,y)t_{rel2}(x,y) * h(x,y;z_{r2}) \propto \delta\left[x + z_{r2}\left(\frac{b}{l_2}+\frac{h}{2(R+d)}-\frac{a}{l_1}\right), y\right], \qquad (2.24)$$

where

$$z_{r2} = \left[\frac{1}{R+d}-\frac{1}{l_1}-\frac{1}{l_2}\right]^{-1} = \frac{(R+d)l_1l_2}{l_1l_2-(l_1+l_2)(R+d)}.$$

Here, z_{r2} is the distance of the real image reconstruction of point object *2* behind the hologram and the image point is located at

$$x = x_2 = -z_{r2}\left(\frac{b}{l_2}+\frac{h}{2(R+d)}-\frac{a}{l_1}\right), \qquad y = y_2 = 0.$$

Equation (2.24) could be obtained alternatively by comparing Eq. (2.23) with (2.19) and noting that we only need to change R to $R + d$ and h to $-h$. The real image reconstructions of point objects 1 and 2 are shown in Fig. 2.6. The locations of the virtual images of point objects 1 and 2 can be similarly calculated starting from Eqs. (2.16c) and (2.16d).

2.3.1 Holographic magnifications

We are now in a position to evaluate the *lateral* and *longitudinal magnifications* of the holographic image and this is best done with the point images we discussed in the last section. The longitudinal distance (along z) between the two real point images is $z_{r2} - z_{r1}$, so the longitudinal magnification is defined as

$$M_{Long}^{r} = \frac{z_{r2} - z_{r1}}{d}. \tag{2.25}$$

Using Eqs. (2.21) and (2.24) and assuming $R \gg d$, the longitudinal magnification becomes

$$M_{Long}^{r} \cong \frac{(l_1 l_2)^2}{(l_1 l_2 - R l_1 - R l_2)^2}. \tag{2.26}$$

We find the lateral distance (along x) between the two image points 1 and 2 by taking the difference between the locations of the two δ-functions in Eqs. (2.22) and (2.24), so the lateral magnification is

$$M_{Lat}^{r} = \frac{z_{r2}\left(\frac{b}{l_2} + \frac{h}{2(R+d)} - \frac{a}{l_1}\right) - z_{r1}\left(\frac{b}{l_2} - \frac{h}{2R} - \frac{a}{l_1}\right)}{h}$$

$$\cong \frac{(z_{r2} - z_{r1})\left(\frac{b}{l_2} - \frac{a}{l_1}\right) + (z_{r2} + z_{r1})\frac{h}{2R}}{h} \tag{2.27}$$

for $R \gg d$. In order to make this magnification independent of the lateral separation between the object points, h, we set

$$\frac{b}{l_2} - \frac{a}{l_1} = 0,$$

or

$$\frac{b}{l_2} = \frac{a}{l_1}. \tag{2.28}$$

Then, from Eq. (2.27) and again for the condition that $R \gg d$,

$$M_{Lat}^{r} \simeq (z_{r2} + z_{r1})\frac{1}{2R} \simeq \frac{l_1 l_2}{l_1 l_2 - (l_1 + l_2)R}. \tag{2.29}$$

By comparing Eq. (2.26) and Eq. (2.29), we have the following important relationship between the magnifications in three-dimensional imaging:

$$M_{Long}^{r} = (M_{Lat}^{r})^2. \tag{2.30}$$

2.3.2 Translational distortion

In the above analysis of magnifications, we have assumed that the condition of Eq. (2.28) is satisfied in the reconstruction. If Eq. (2.28) is violated, the lateral magnification will depend on the lateral separation between the object points. In other words, the reconstructed image will experience *translational distortion*. To see clearly the effect of translational distortion, let us consider point objects 1 and 2 along the z-axis by taking $h = 0$ and inspecting their image reconstruction locations. The situation is shown in Fig. 2.7. Points 1 and 2 are shown in the figure as a reference to show the original image locations. Points $1'$ and $2'$ are reconstructed real image points of object points 1 and 2, respectively, due to the reconstruction wave from point r. We notice there is a translation between the two real image points along the x-direction. The translational distance Δx is given by

$$\Delta x = x_1 - x_2,$$

where x_1 and x_2 are the locations previously found [see below Eqs. (2.22) and (2.24)]. For $h = 0$, we find the translational distance

$$\Delta x = (z_{r2} - z_{r1})\left(\frac{b}{l_2} - \frac{a}{l_1}\right). \tag{2.31}$$

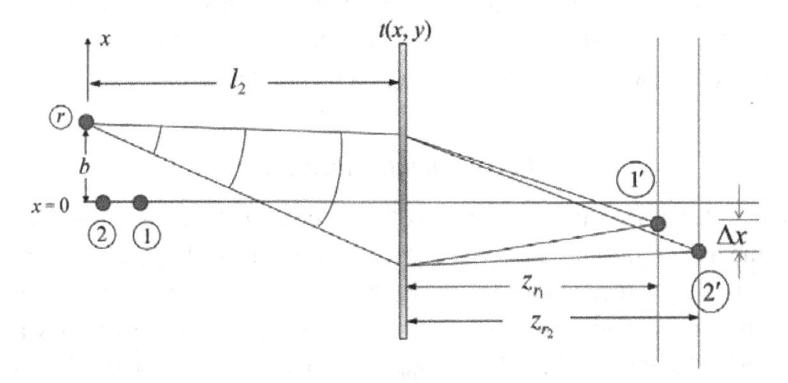

Figure 2.7 Translational distortion of the reconstructed real image.

From the above result, we see that the image is twisted for a three-dimensional object. In practice we can remove the translational distortion by setting the reconstruction point to satisfy Eq. (2.28). The distortion can also be removed by setting $a = b = 0$. However, by doing so, we lose the separation of the real, virtual, and zeroth-order diffraction, a situation reminiscent of what we observed in Fig. 2.3 where we used plane waves for recording and reconstruction, with both the plane waves traveling along the same direction. We will discuss this aspect more in Chapter 3.

Example 2.1: Holographic magnification

In this example, we show that by using a spherical wave for recording and plane wave reconstruction, we can produce a simple magnification imaging system. We start with a general result for M^r_{Lat} given by Eq. (2.27):

$$M^r_{Lat} = \frac{z_{r2}\left(\frac{b}{l_2} + \frac{h}{2(R+d)} - \frac{a}{l_1}\right) - z_{r1}\left(\frac{b}{l_2} - \frac{h}{2R} - \frac{a}{l_1}\right)}{h}. \tag{2.32}$$

For $a = b = 0$, i.e., the recording and reconstruction point sources are on the z-axis, and $d = 0$, i.e., we are considering a planar image, M^r_{Lat} becomes

$$M^r_{Lat} = \frac{z_{r2} + z_{r1}}{2R} = \left[1 - \frac{R}{l_1} - \frac{R}{l_2}\right]^{-1}, \tag{2.33}$$

where $z_{r2} = z_{r1} = [1/R - 1/l_1 - 1/l_2]^{-1}$. For plane wave reconstruction, $l_2 \to \infty$. Equation (2.33) finally becomes a simple expression given by

$$M^r_{Lat} = \left[1 - \frac{R}{l_1}\right]^{-1}. \tag{2.34}$$

For example, taking $l_1 = 2R$, $M^r_{Lat} = 2$, a magnification of a factor of 2, and for $l_1 = R/4 < R$, $M^r_{Lat} = -1/3$, a demagnification in this case. Note that if the recording reference beam is also a plane wave, i.e., $l_1 \to \infty$, there is no magnification using a plane wave for recording and reconstruction.

2.3.3 Chromatic aberration

In the above discussion, the wavelength of the reconstruction wave was assumed to be the same as that of the wave used for holographic recording. If the hologram is illuminated using a reconstruction wave with a different wavelength, λ_r, then the situation becomes much more complicated. Now the reconstruction wave can still be found using Eq. (2.19) but $\psi_{pr}(x, y)$ and $h(x, y; z)$ must be modified according to

$$\psi_{pr}(x,y) \propto \exp\left[\frac{-jk_r}{2l_2}\left((x-b)^2+y^2\right)\right],$$

and

$$h(x,y;z) \propto \frac{jk_r}{2\pi z}\exp\left[\frac{-jk_r}{2z}(x^2+y^2)\right],$$

respectively, where $k_r = 2\pi/\lambda_r$. Hence the longitudinal distance of the real image reconstruction [see Eq. (2.21)] is modified to become

$$z_{r1} = \left[\frac{\lambda_r}{\lambda_0 R}-\frac{\lambda_r}{\lambda_0 l_1}-\frac{1}{l_2}\right]^{-1} = \frac{\lambda_0 R l_1 l_2}{\lambda_r l_1 l_2-(\lambda_r l_2+\lambda_0 l_1)R}. \tag{2.35}$$

Accordingly, the transverse location can be found from Eq. (2.22) to give

$$x_1 = -z_{r1}\left(\frac{b}{l_2}-\frac{h\lambda_r}{2R\lambda_0}-\frac{a\lambda_r}{l_1\lambda_0}\right), \qquad y_1 = 0. \tag{2.36}$$

Thus in general the location of the image point depends on the wavelength of the reconstruction wave, resulting in chromatic aberration. We can see that for $R \ll l_1$ and $R \ll l_2$,

$$z_{r1} \approx \frac{\lambda_0}{\lambda_r}R, \tag{2.37a}$$

$$x_1 \approx \frac{h}{2}+R\left(\frac{a}{l_1}-\frac{b}{l_2}\frac{\lambda_0}{\lambda_r}\right). \tag{2.37b}$$

As a result, in chromatic aberration, the shift of the image location due to the difference of the wavelength used for recording and reconstruction is proportional to R, the distance of the object from the hologram.

Example 2.2: Chromatic aberration calculation

We calculate the chromatic aberration of an image point in the following case: $R = 5$ cm, $h = 2$ cm, $a = b = 5$ cm, $l_1 = l_2 = 20$ cm, $\lambda_0 = 632$ nm.

We define the longitudinal aberration distance, δz, and the transverse aberration distance, δx, as

$$\delta z = z_{r1}(\lambda_r) - z_{r1}(\lambda_0), \tag{2.38a}$$

$$\delta x = x_1(\lambda_r) - x_1(\lambda_0). \tag{2.38b}$$

δz and δx are plotted in Fig. 2.8 with the MATLAB code listed in Table 2.1. In comparison with the desired image point, $z_{r1}(\lambda_0) = 10$ cm and $x_1(\lambda_0) = 2$ cm, the amount of aberration increases as the deviation from the desired wavelength, λ_0, becomes larger so that holograms are usually reconstructed with a single

Table 2.1 *MATLAB code for chromatic aberration calculation, see Fig. 2.8*

```
close all; clear all;
L=20;  % L1 and L2
R=5;
a=5;   % a and b
h=2;
lambda0=633;        %recording wavelength
lambdaR=400:20:700; %reconstruction wavelength
z=lambda0*R*L./(lambdaR*(L-R)-lambda0*R);
dz=R*L/(L-2*R)-z;
plot(lambdaR,dz)
title('Longitudinal chromatic aberration')
xlabel('Reconstruction wavelength (nm)')
ylabel('{\delta} z (mm)')
x=-z.*(a/L-lambdaR/lambda0*h/R/2-lambdaR*a/lambda0/L);
dx=x-R*L/(L-2*R)*(h/2/R);
figure;plot(lambdaR,dx)
title('Transverse chromatic aberration')
xlabel('Reconstruction wavelength (nm)')
ylabel('{\delta} x (mm)')
```

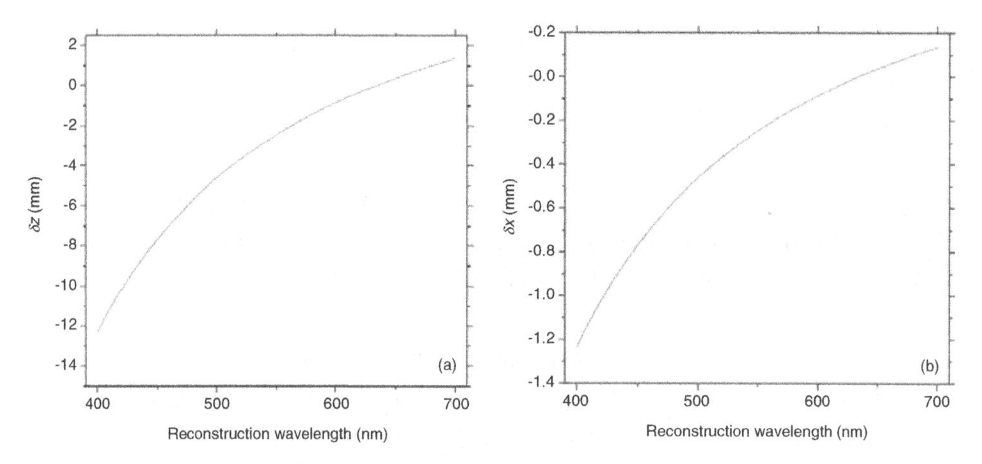

Figure 2.8 (a) Longitudinal, and (b) transverse chromatic aberration distances when the recording wavelength is $\lambda_0 = 632$ nm.

wavelength. In the next chapter we will see that holograms can be reconstructed using white light in some specific kinds of geometries.

2.4 Temporal and spatial coherence

In the preceding discussions when we have discussed holographic recording, we have assumed that the optical fields are completely coherent and monochromatic so that the fields will always produce interference. In this section, we give a brief introduction to temporal and spatial coherence. In temporal coherence, we are concerned with the ability of a light field to interfere with a time-delayed version of itself. In spatial coherence, the ability of a light field to interfere with a spatially shifted version of itself is considered.

2.4.1 Temporal coherence

In a simplified analysis of interference, light is considered to be monochromatic, i.e., the bandwidth of the light source is infinitesimal. In practice there is no ideal monochromatic light source. A real light source contains a range of frequencies and hence interference fringes do not always occur. An interferogram is a photographic record of intensity versus optical path difference of two interfering waves. The interferogram of two light waves at r is expressed as

$$\begin{aligned} I &= \left\langle |A(r,t) + B(r,t)|^2 \right\rangle \\ &= \left\langle |A(r,t)|^2 \right\rangle + \left\langle |B(r,t)|^2 \right\rangle + 2 \times \mathrm{Re}\left\{ \langle A^*(r,t) B(r,t) \rangle \right\}, \end{aligned} \tag{2.39}$$

where $\langle \cdot \rangle$ stands for the time-average integral as

$$\langle \cdot \rangle = \lim_{T \to \infty} \frac{1}{\mathrm{T}} \int_{-T/2}^{T/2} \cdot \, dt, \tag{2.40}$$

and $A(r, t)$ and $B(r, t)$ denote the optical fields to be superimposed. In the following discussion we will first assume that the two light fields are from an infinitesimal, quasi-monochromatic light source. We model the quasi-monochromatic light as having a specific frequency ω_0 for a certain time and then we change its phase randomly. Thus at fixed r, $A(t)$ and $B(t)$ can be simply expressed as

$$A(t) = A_0 \exp\{ j[\omega_0 t + \theta(t)] \}, \tag{2.41a}$$

$$B(t) = B_0 \exp\{ j[\omega_0 (t + \tau) + \theta(t + \tau)] \}, \tag{2.41b}$$

where τ denotes the time delay due to the optical path difference between $A(t)$ and $B(t)$, and $\theta(t)$ denotes the time-variant initial phase of the quasi-monochromatic light. By substituting Eq. (2.41) into Eq. (2.39), we have

$$I(\tau) = A_0^2 + B_0^2 + 2A_0 B_0 \times \mathrm{Re}\left\{ \left\langle e^{j[\theta(t+\tau) - \theta(t) + \omega_0 \tau]} \right\rangle \right\} \tag{2.42}$$

because $\langle |A(t)|^2 \rangle = A_0^2$ and $\langle |B(t)|^2 \rangle = B_0^2$. In Eq. (2.42), the time-average integral is the interference term called the *complex degree of coherence* of the source, which is denoted as

$$\gamma(\tau) = \left\langle e^{j[\theta(t+\tau) - \theta(t) + \omega_0 \tau]} \right\rangle. \tag{2.43}$$

The complex degree of coherence has the properties

$$\gamma(0) = 1 \qquad \text{and} \qquad |\gamma(\tau)| \leq 1. \tag{2.44}$$

As a result, the interferogram can be expressed in terms of the complex degree of coherence as

$$I(\tau) = A_0^2 + B_0^2 + 2A_0B_0|\gamma(\tau)|\cos[\arg\{\gamma(\tau)\}], \tag{2.45}$$

where $\arg\{\bullet\}$ stands for the operation of taking the argument of the function being bracketed. It should be noted that in Eq. (2.45) the modulus of the complex degree of coherence comes into existence only when we measure the intensity and it is not directly obtainable. In fact, the modulus of the complex degree of coherence is easy to determine by measuring the contrast between fringes in $I(\tau)$, as first performed by Michelson. The fringe contrast is called the *fringe visibility* V, defined by

$$V = \frac{I_{max} - I_{min}}{I_{max} + I_{min}}, \tag{2.46}$$

where I_{max} and I_{min} denote the local maximum value and the local minimum value of the interferogram, respectively. Accordingly, we can see that

$$I_{max} = A_0^2 + B_0^2 + 2A_0B_0|\gamma(\tau)|,$$
$$I_{min} = A_0^2 + B_0^2 - 2A_0B_0|\gamma(\tau)|.$$

So the visibility of the interferogram in Eq. (2.46) can be expressed as

$$V = \frac{2A_0B_0}{A_0^2 + B_0^2}|\gamma(\tau)|. \tag{2.47}$$

Equation (2.47) shows that the modulus of the degree of coherence is proportional to the visibility of the fringe. So we can deduce the ability to form interference from a light source if we know its coherence property. We say that light waves involved in an interferometer are completely coherent, completely incoherent, or partially coherent according to the value of $|\gamma(\tau)|$:

$$|\gamma(\tau)| = 1 \quad \text{complete coherence}$$
$$|\gamma(\tau)| = 0 \quad \text{complete incoherence}$$
$$0 \leq |\gamma(\tau)| \leq 1 \quad \text{partial coherence.}$$

Let us take a simple plane wave as an example, i.e., $A(t) = A_0 \exp(j\omega_0 t)$, and $B(t) = A_0 \exp[j\omega_0(t + \tau)]$. Equation (2.43) becomes $\gamma(\tau) = \exp(j\omega_0\tau)$ and therefore $|\gamma(\tau)| = 1$, a case of complete coherence. On the other hand, if $A(t)$ is completely random in time, from Eq. (2.43), we have $\gamma(\tau) = 0$, a case of complete incoherence.

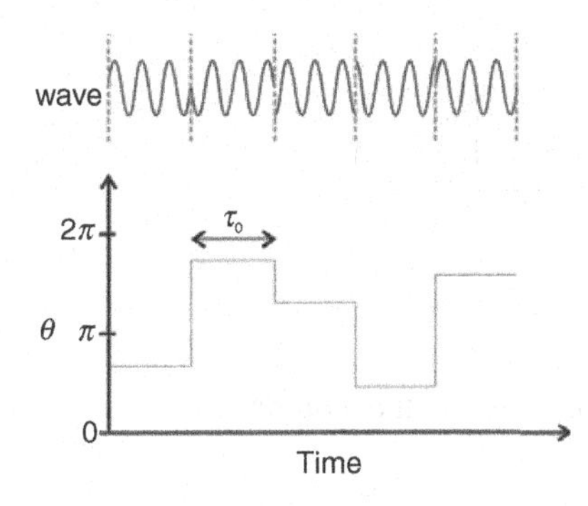

Figure 2.9 Phase function $\theta(t)$ for a quasi-monochromatic wave, showing random phase fluctuations every τ_0 of the wave (shown at the top of the figure).

Many natural and artificial light sources have a monotonously decreasing function in $|\gamma(\tau)|$, starting from $|\gamma(0)| = 1$.

2.4.2 Coherence time and coherence length

When we consider interference and diffraction of waves, we assume that the light field remains perfectly sinusoidal for all time. But this idealized situation is not true for ordinary light sources. We can model an ordinary light source as a quasi-monochromatic light oscillating at ω_0 of finite size wave trains with initial phase $\theta(t)$ to be randomly distributed between 0 and 2π within some fixed time, i.e., the phase changes randomly every time interval τ_0 and remains stable between the changes, as shown in Fig. 2.9. According to the model, the complex degree of coherence can be found by evaluating Eq. (2.43) to be

$$\gamma(\tau) = \Lambda\left(\frac{\tau}{\tau_0}\right)e^{j\omega_0\tau}, \tag{2.48}$$

where $\Lambda(\tau/\tau_0)$ is a triangle function as defined in Table 1.1 and is repeated below for convenience:

$$\Lambda\left(\frac{\tau}{\tau_0}\right) = \begin{cases} 1 - \left|\dfrac{\tau}{\tau_0}\right| & \text{for}\left|\dfrac{\tau}{\tau_0}\right| \leq 1 \\ 0 & \text{otherwise.} \end{cases}$$

The modulus of $\gamma(\tau)$ is plotted in Fig. 2.10. It is shown that $|\gamma(\tau)|$ decreases with τ and falls to zero when $\tau \geq \tau_0$. By substituting Eq. (2.48) into Eq. (2.45), the interferogram can be expressed as

$$\begin{aligned} I(\tau) &= A_0^2 + B_0^2 + 2A_0B_0\Lambda\left(\frac{\tau}{\tau_0}\right)\cos\left[\omega_0\tau\right] \\ &= A_0^2 + B_0^2 + 2A_0B_0\Lambda\left(\frac{\tau}{\tau_0}\right)\cos\left[2\pi\Delta d/\lambda_0\right], \end{aligned} \tag{2.49}$$

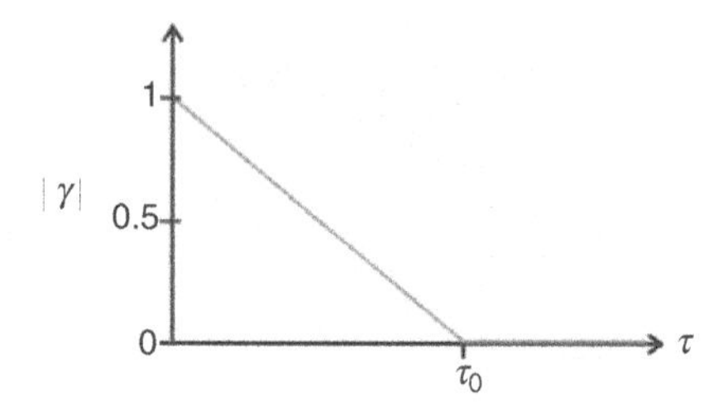

Figure 2.10 The modulus of the complex degree of coherence for quasi-monochromatic light.

where Δd is the optical path difference corresponding to the time delay τ between the two light waves, i.e., $2\pi\Delta d/\lambda_0 = \omega_0\tau$.

The width of the complex degree of coherence, τ_0, is called the *coherence time*. If the time delay between the light waves involved in the interference is larger than the coherence time, no fringes can be observed.

Finally, we can also define the *coherence length* ℓ_c as

$$\ell_c = c\tau_0, \tag{2.50}$$

where c is the speed of light in vacuum. In other words, the coherence length is the path the light passes in the time interval τ_0. To ensure the success of interference, the optical path difference in an interferometer must be smaller than the coherence length.

2.4.3 Some general temporal coherence considerations

In the above discussion, we used the model of a quasi-monochromatic light so that the analysis is relatively simple. Here we will extend the theory to any kind of light source. In Eq. (2.42), the complex degree of coherence comes from the process of time average of the cross term. On the other hand, we also know that $\gamma(0) = 1$ according to Eq. (2.44). Thus for any light source, we can write down the complex degree of coherence as:

$$\gamma(\tau) = \frac{\langle E^*(t)E(t+\tau)\rangle}{\langle |E(t)|^2\rangle}, \tag{2.51}$$

where $E(t)$ is the complex amplitude of the light at the source point. Equation (2.51) is the general form of the complex degree of coherence. It should be noted that the frequency range of the optical wave $E(t)$ cannot be detected directly using any modern photodetectors. Hence, to evaluate Eq. (2.51), we first express the time average $\langle E^*(t)E(t+\tau)\rangle$ as an auto-correlation calculation,

$$\langle E^*(t)E(t+\tau)\rangle = \lim_{T\to\infty}\frac{1}{T}\int_{-T/2}^{T/2} E^*(t)E(t+\tau)\,dt$$
$$\propto [E(\tau)\otimes E(\tau)]. \tag{2.52}$$

From the correlation of the Fourier transform in Table 1.1, we have

$$\mathcal{F}_\tau\{E(\tau)\otimes E(\tau)\} = \int_{-\infty}^{\infty}[E(\tau)\otimes E(\tau)]\exp(-j\omega\tau)d\tau = |\mathcal{F}_\tau\{E(\tau)\}|^2 = \mathcal{P}_c(\omega), \tag{2.53}$$

where we define the one-dimensional temporal Fourier transform of $f(\tau)$ as

$$\mathcal{F}_\tau\{f(\tau)\} = \int_{-\infty}^{\infty} f(\tau)\exp(-j\omega\tau)d\tau = F(\omega), \tag{2.54a}$$

and the inverse temporal Fourier transform is

$$\mathcal{F}_\tau^{-1}\{F(\omega)\} = \frac{1}{2\pi}\int_{-\infty}^{\infty} F(\omega)\exp(j\omega\tau)d\omega = f(\tau), \tag{2.54b}$$

where τ and ω are the time and the temporal radian frequency variables, respectively. Note that the definitions of $\mathcal{F}_\tau$ and $\mathcal{F}_\tau^{-1}$ are different from those of $\mathcal{F}$ and $\mathcal{F}^{-1}$ for the spatial function defined in Eq. (1.22); namely "$-j$" is used in the exponential function for the time function in the forward transform. This is done purposely to be consistent with the engineering convention for a traveling wave. In the convention, $\exp[j(\omega_0 t - k_0 z)]$ denotes a plane wave traveling in the $+z$-direction. However, we can still use Table 1.1 as long as we replace "j" by "$-j$" in the transform pairs.

Let us return our attention to Eq. (2.53). The result of Eq. (2.53) is the *Wiener–Khinchin theorem*, which states that the Fourier transform of the auto-correlation of the light field is proportional to the power spectrum of the light source, $\mathcal{P}_c(\omega)$. $\mathcal{P}_c(\omega)$ is the power spectrum of the complex field E, i.e., a field represented by a complex quantity. It should be noted that $\mathcal{P}_c(\omega)$ is not the same as $\mathcal{P}_r(\omega)$, the power spectrum of the real light field where the light field is expressed in terms of a real quality. $\mathcal{P}_r(\omega)$ includes both $+\omega$ and $-\omega$ components, while $\mathcal{P}_c(\omega)$ contains only the $+\omega$ component. As it turns out, the relation between $\mathcal{P}_c(\omega)$ and $\mathcal{P}_r(\omega)$ is simple, namely [4]

$$\mathcal{P}_c(\omega) = \begin{cases} 4\mathcal{P}_r(\omega) & \text{for } \omega > 0 \\ 0 & \text{for } \omega < 0. \end{cases}$$

As a result, Eq. (2.51) can be re-written to become

$$\gamma(\tau) = \frac{\mathcal{F}_\tau^{-1}\{\mathcal{P}_c(\omega)\}}{\mathcal{F}_\tau^{-1}\{\mathcal{P}_c(\omega)\}_{\tau=0}} = \mathcal{F}_\tau^{-1}\{\mathcal{P}_N(\omega)\}, \tag{2.55}$$

where $\mathcal{P}_N(\omega)$ is the normalized power spectrum of the source and it ensures $\gamma(0) = 1$.

2.4.4 Fourier transform spectroscopy

In practice, the power spectrum can be deduced from measurements of $I(\tau)$. We let $A(t) = A_0(\omega)\, e^{j\omega t}$, a plane wave of frequency ω with its amplitude given by $A_0(\omega)$, and $B(t) = A_0(\omega)\, e^{j\omega(t+\tau)}$ as $B(t)$ is the time-delayed version of the same point from the source. According to Eq. (2.39),

$$I(\tau) = \langle |A_0(\omega)e^{j\omega t} + A_0(\omega)e^{j\omega(t+\tau)}|^2\rangle = 2|A_0(\omega)|^2[1 + \cos(\omega\tau)]. \tag{2.56}$$

Now we take the source which is composed of a collection of waves with different frequencies as follows:

$$E(t) = \int_0^\infty A_0(\omega)\exp(j\omega t)d\omega. \tag{2.57}$$

The total contribution from all the frequencies to $I(\tau)$ then becomes

$$\begin{aligned} I(\tau) &= \int_0^\infty 2|A_0(\omega)|^2[1 + \cos(\omega\tau)]d\omega \\ &\propto 2\int_0^\infty \mathcal{P}_r(\omega)d\omega + 2\int_0^\infty \mathcal{P}_r(\omega)\cos(\omega\tau)d\omega \\ &\propto 2\, I_0 + \Delta I(\tau). \end{aligned} \tag{2.58}$$

As $|A_0(\omega)|^2 \propto \mathcal{P}_r(\omega)$, the power spectrum of the source, the first term in the above equation is a constant and simply proportional to the total intensity of the light source, I_0. The second term $\Delta I(\tau)$ varies with the delay τ. Note that the power spectrum is only given for positive frequencies and by letting $\mathcal{P}_r(-\omega) = \mathcal{P}_r(\omega)$, an even function in ω, we can extend the second term of the equation to become

$$\begin{aligned} \Delta I(\tau) &= 2\int_0^\infty \mathcal{P}_r(\omega)\cos(\omega\tau)d\omega = \int_{-\infty}^\infty \mathcal{P}_r(\omega)\exp(j\omega\tau)d\omega \\ &\propto \frac{1}{2\pi}\int_{-\infty}^\infty \mathcal{P}_r(\omega)\exp(j\omega\tau)d\omega = \mathcal{F}_\tau^{-1}\{\mathcal{P}_r(\omega)\}. \end{aligned} \tag{2.59}$$

By taking the temporal Fourier transform of the above equation, we have

$$\mathcal{F}_\tau\{\Delta I(\tau)\} = \mathcal{P}_r(\omega). \tag{2.60}$$

The above equation is an important result. Once all the measurements of $I(\tau)$ are collected for different time delay τ, the Fourier transform is calculated to give the power spectrum of the light – such a measurement technique is called *Fourier transform spectroscopy*.

We now want to relate some power spectra to $\gamma(\tau)$. Starting from Eq. (2.39), we let $A(r, t) = E(t)$ and $B(r, t) = E(t + \tau)$. We can write $I(\tau)$ as

$$\begin{aligned} I(\tau) = \left\langle |E(t) + E(t+\tau)|^2 \right\rangle &= 2\left\langle |E(t)|^2 \right\rangle + 2\mathrm{Re}\left\{ \langle E^*(t)E(t+\tau) \rangle \right\} \\ &= 2\left\langle |E(t)|^2 \right\rangle \left[1 + \mathrm{Re}\{\gamma(\tau)\} \right], \end{aligned} \tag{2.61}$$

where we have used Eq. (2.45) to obtain the last step of the above equation. Using the model of a partially coherent light as a quasi-monochromatic light of finite size wave trains with random initial phase, we have, from Eq. (2.48), $\gamma(\tau) = \Lambda(\tau/\tau_0)e^{j\omega_0\tau}$, and Eq. (2.61) becomes

$$I(\tau) = 2\left\langle |E(t)|^2 \right\rangle [1 + |\gamma(\tau)| \cos(\omega_0\tau)] = 2\left\langle |E(t)|^2 \right\rangle \left[1 + \Lambda\left(\frac{\tau}{\tau_0}\right) \cos(\omega_0\tau) \right]. \tag{2.62}$$

The power spectrum, according to Eq. (2.55), is

$$\mathcal{P}_N(\omega) \propto \mathcal{F}_\tau\{\gamma(\tau)\} = \mathcal{F}_\tau\left\{ \Lambda\left(\frac{\tau}{\tau_0}\right) e^{j\omega_0\tau} \right\} = \tau_0 \,\mathrm{sinc}^2\left[\frac{(\omega - \omega_0)\tau_0}{2\pi} \right], \tag{2.63}$$

where we have used the transform pairs in Table 1.1, taking into account that we are dealing with one-dimensional time functions. Note that the full width at half-maximum (FWHM) of the power spectrum $\Delta\omega$ is related to τ_0 by $\tau_0 = 5.566/\Delta\omega$. Other examples of spectra are the Gaussian spectrum and the rectangular spectrum. For a Gaussian spectrum, we have

$$\mathcal{P}_N(\omega) \propto \exp\left[-\left(\frac{\omega - \omega_0}{2\sigma^2} \right)^2 \right] \tag{2.64a}$$

with its complex degree of coherence given by

$$\gamma(\tau) \propto \exp\left(-\frac{\sigma^2\tau^2}{2} \right) e^{j\omega_0\tau}, \tag{2.64b}$$

where the FWHM of the power spectrum $\Delta\omega$ is related to σ by $\sigma = \Delta\omega/\sqrt{8 \times \ln 2}$. For a rectangular spectrum, we have

$$\mathcal{P}_N(\omega) \propto \mathrm{rect}\left(\frac{\omega - \omega_0}{\Delta\omega} \right) \tag{2.65a}$$

and its complex degree of coherence is

$$\gamma(\tau) \propto \text{sinc}\left(\frac{\Delta\omega\tau}{2\pi}\right)e^{j\omega_0\tau}. \tag{2.65b}$$

Although we have indicated that the coherence time is the width of $|\gamma(\tau)|$, for most practical light sources it is hard to determine the coherence time by intuition. So here we adopt the definition of coherence time τ_c proposed by Mandel [5], that is

$$\tau_c \equiv \int_{-\infty}^{\infty} |\gamma(\tau)|^2 d\tau. \tag{2.66}$$

For example, the coherence time of a quasi-monochromatic light is found to be $\tau_c = 2\tau_0/3$ according to the definition in Eq. (2.66). Finally, the coherence time always has a value of the same order of magnitude as the reciprocal of the bandwidth of the source, that is

$$\tau_c \sim \frac{2\pi}{\Delta\omega}. \tag{2.67}$$

As a result, we can also show that, using Eqs. (2.50) and (2.67) and taking $\tau_c \sim \tau_0$, we have

$$\ell_c \sim \frac{\lambda_0^2}{\Delta\lambda}, \tag{2.68}$$

where $\Delta\lambda$ is the spectral line width of the source and λ_0 is the wavelength corresponding to the center frequency of the power spectrum ω_0. We can also relate the spectral line width to the FWHM of the power spectrum $\Delta\omega$ as follows:

$$\Delta\omega \sim \frac{2\pi c}{\lambda_0^2}\Delta\lambda. \tag{2.69}$$

White light has a line width of about 300 nm, ranging roughly from 400 to 700 nm, and if we take the average wavelength at 550 nm, Eq. (2.68) gives $\ell_c \sim 1$ μm, a very short coherence length. LEDs have a spectral width $\Delta\lambda$ of about 50 nm and have a coherence length of about 7 μm for red color with wavelength of 0.6 μm. As for the green line of mercury at 546 nm having a line width about 0.025 nm, its coherence length is about 1.2 cm. Lasers typically have a long coherence length. Helium–neon lasers can produce light with coherence lengths greater than 5 m, but 20 cm is typical. Some industrial CO_2 lasers of line width of around 10^{-5} nm producing emission at the infrared wavelength of 10.6 μm would give a coherence length of around 11 km.

2.4.5 Spatial coherence

In the above discussion of temporal coherence, light was assumed to be emitted from an infinitesimal source and we have found the ability of two relatively delayed light waves to form fringes. In practice, any light source must have a size of extended area, and light emitted from any two points of this area is incoherent. In this subsection, the dependence of the finite size of the quasi-monochromatic incoherent light source on interference is discussed, which brings us to the concept of spatial coherence.

To simplify the analysis, we first consider interference including two identical but incoherent quasi-monochromatic light sources, as shown in Fig. 2.11. If there is a single point source, say source 1, light emitted from the source is separated into two paths, one path passing through point A and the other path through point B. Light emerging from points A and B comes to the screen to form interference fringes. Hence we may think of A and B as two slits in this situation. The corresponding analysis is similar to that in the above subsection. Now we will deal with the case that two point sources, source 1 and source 2, exist at the same time.

In the plane of the screen where we have interference, we only consider point P, the center of the screen. Thus the optical path difference between $\overline{\mathrm{AP}}$ and $\overline{\mathrm{BP}}$ is zero. At point P, we have

$$\begin{aligned} I &= \left\langle |E_A(t) + E_B(t)|^2 \right\rangle \\ &= \left\langle |E_A(t)|^2 \right\rangle + \left\langle |E_B(t)|^2 \right\rangle + 2 \times \mathrm{Re}\left\{ \langle E_A{}^*(t) E_B(t) \rangle \right\}, \end{aligned} \tag{2.70}$$

where

$$E_A(t) = E_{1A}(t) + E_{2A}(t), \tag{2.71a}$$

and

$$E_B(t) = E_{1B}(t) + E_{2B}(t) \tag{2.71b}$$

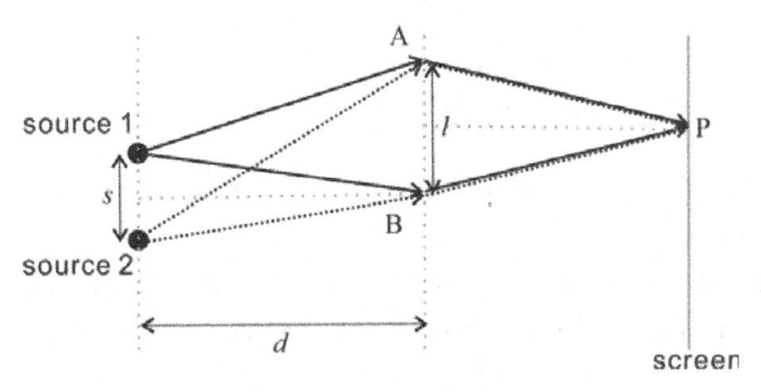

Figure 2.11 Schematic diagram of interference formed by two incoherent light sources.

are the optical fields passing through point A and point B, respectively; $E_{1i}(t)$ and $E_{2i}(t)$ are the components emitted from source 1 and source 2, respectively, where the subscript $i = A$ or B. According to the concept of temporal coherence [Eq. (2.51)], we can define the complex degree of coherence between points A and B, that is

$$\gamma_{AB} = \frac{\langle E_A^*(t)E_B(t)\rangle}{\sqrt{\langle |E_A(t)|^2\rangle\langle |E_B(t)|^2\rangle}}, \tag{2.72}$$

which is called the *complex degree of mutual coherence*, and γ of Eq. (2.51) is, strictly speaking, the *complex degree of self coherence*.

Because source 1 and source 2 are completely incoherent, they cannot interfere with each other, i.e., cross terms such as $\langle E_{1A}^*(t)E_{2B}(t)\rangle = 0$, $\langle E_{2A}^*(t)E_{1B}(t)\rangle = 0$, etc. Thus we have the following results:

$$\begin{aligned}\langle E_A^*(t)E_B(t)\rangle &= \langle E_{1A}^*(t)E_{1B}(t)\rangle + \langle E_{2A}^*(t)E_{2B}(t)\rangle,\\ \langle |E_A|^2\rangle &= \langle |E_{1A}|^2\rangle + \langle |E_{2A}|^2\rangle,\\ \langle |E_B|^2\rangle &= \langle |E_{1B}|^2\rangle + \langle |E_{2B}|^2\rangle.\end{aligned}$$

Suppose that $\langle |E_{1A}|^2\rangle \approx \langle |E_{1B}|^2\rangle \approx \langle |E_{2A}|^2\rangle \approx \langle |E_{2B}|^2\rangle$ for simplicity, then we can rewrite Eq. (2.72) as

$$\gamma_{AB} = \frac{\langle E_{1A}^*(t)E_{1B}(t)\rangle}{2\langle |E_{1A}(t)|^2\rangle} + \frac{\langle E_{2A}^*(t)E_{2B}(t)\rangle}{2\langle |E_{2A}(t)|^2\rangle}. \tag{2.73}$$

Because both $E_{1A}(t)$ and $E_{1B}(t)$ come from the same source, their relationship must be $E_{1B}(t) = E_{1A}(t + \tau_1)$; similarly, we will also have $E_{2B}(t) = E_{2A}(t + \tau_2)$, where τ_1 and τ_2 are the time delays between the waves at point A and point B, respectively, from source 1 and source 2. As a result, the complex degree of coherence γ_{AB} can be expressed as

$$\begin{aligned}\gamma_{AB} &= \frac{\langle E_{1A}^*(t)E_{1A}(t+\tau_1)\rangle}{2\langle |E_{1A}(t)|^2\rangle} + \frac{\langle E_{2A}^*(t)E_{2A}(t+\tau_2)\rangle}{2\langle |E_{2A}(t)|^2\rangle}\\ &= \frac{1}{2}\gamma_1(\tau_1) + \frac{1}{2}\gamma_2(\tau_2).\end{aligned} \tag{2.74}$$

Equation (2.74) shows that γ_{AB} depends on the complex degree of self coherence of the two source points.

If the light emitted from the two point sources is quasi-monochromatic, we can use the result from Eq. (2.48) for Eq. (2.74) to yield

$$\gamma_{AB} = \frac{1}{2}\Lambda\left(\frac{\tau_1}{\tau_0}\right)e^{j\omega\tau_1} + \frac{1}{2}\Lambda\left(\frac{\tau_2}{\tau_0}\right)e^{j\omega\tau_2}. \tag{2.75}$$

After some manipulation, the modulus of the complex degree of coherence can be expressed as

$$|\gamma_{AB}|^2 \approx \frac{1}{2}[1 + \cos\omega(\tau_1 - \tau_2)]\Lambda\left(\frac{\tau_1}{\tau_0}\right)\Lambda\left(\frac{\tau_2}{\tau_0}\right) \tag{2.76}$$

provided that $\tau_1 \approx \tau_2 \gg |\tau_1 - \tau_2|$. Hence the period of the cosine function in Eq. (2.76) is much shorter than that of the triangle function and the cosine function dominates the visibility of the interferogram. By applying the simple geometry described in Fig. 2.11 and using the paraxial approximation, we can obtain

$$\tau_2 - \tau_1 = \frac{sl}{dc}, \tag{2.77}$$

where s is the separation of the two light sources, l is the separation of points A and B, and d is the separation of the source plane from the plane of points A and B. Finally, we can say that the interference fringes are visible, provided

$$\omega|\tau_1 - \tau_2| < \pi,$$

or

$$l < \frac{d\lambda_0}{2s}. \tag{2.78}$$

To understand the meaning of Eq. (2.78), we can imagine that in Fig. 2.11 point A is a movable pinhole while point B is a fixed pinhole. When the separation between point A and point B is very short, i.e., $l \ll d\lambda_0/2s$, the visibility of the interferogram on the screen approaches unity. When the separation between point A and point B is $d\lambda_0/2s$, the visibility falls to zero. As a result, we can define the *transverse coherence length* l_t as

$$l_t = \frac{d\lambda_0}{2s} = \frac{\lambda_0}{2\theta_s}, \tag{2.79}$$

where $\theta_s = s/d$ is the angular separation of the source measured from the plane of points A and B. In other words, we can improve the spatial coherence by moving the source far away from the plane of points A and B.

2.4.6 Some general spatial coherence considerations

In the above discussion, the quasi-monochromatic incoherent light source only includes two points. This is, of course, not the real situation. In practice, the source is extended and consists of numerous independent point sources. Now we want to know the correlation between a fixed point B and another fixed point A. As shown in Fig. 2.12(a), every point source S in the source emits light to both point A and

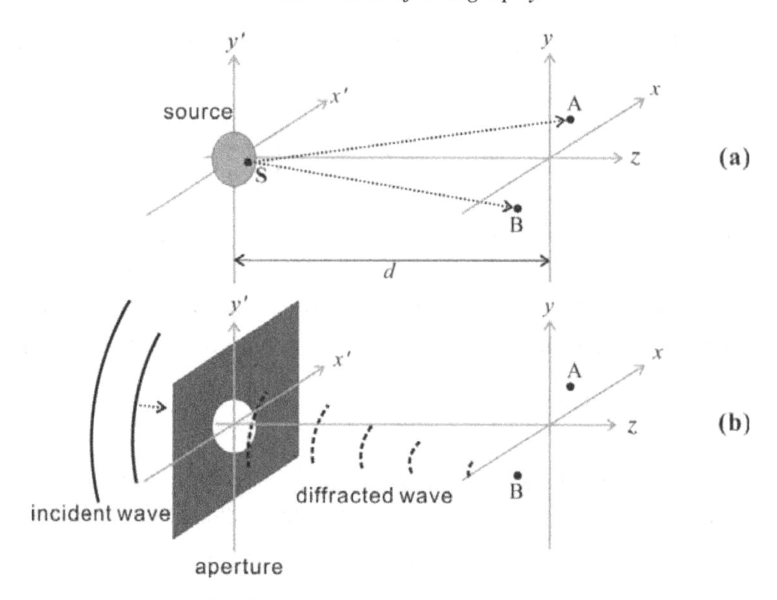

Figure 2.12 Schematic diagrams of (a) correlation between light of two points A and B from an extended source, and (b) corresponding calculation model.

point B, and thus there is a specific correlation between A and B. Nevertheless, we must take into account all point sources in the source to find the overall correlation between A and B, and the analysis becomes complicated. To simplify the analysis, we can make use of the *Van Cittert–Zernike theorem*. The theorem states that the problem of spatial coherence shown in Fig. 2.12(a) can be evaluated by solving the diffraction problem shown in Fig. 2.12(b). First the extended source is replaced by an aperture with the same amplitude distribution proportional to the source, $S(x, y)$. The aperture is then illuminated by a spherical wave converging to a fixed point B. The diffracted field can be found using the Fresnel diffraction formula [Eq. (1.35)]. Specifically, the optical field on the (x, y) plane can be regarded as a diffraction-limit focused point centered at B, and is given by

$$\psi_p(x,y;x_b,y_b) = C \times \exp\left[\frac{-jk_0}{2d}(x^2+y^2)\right]$$
$$\times \iint_{-\infty}^{\infty} S(x',y')\exp\left\{\frac{+jk_0}{d}[(x-x_b)x'+(y-y_b)y']\right\}dx'\,dy', \quad (2.80)$$

where C is a proportionality constant, and (x_b, y_b) is the location of point B. Finally, the complex degree of coherence between points A and B is in the form of the normalized diffracted field. Explicitly, the complex degree of coherence can be calculated as

$$\gamma_{AB} = \frac{\iint_{-\infty}^{\infty} S(x', y') \exp\left\{ \frac{+jk_0}{d} [(x_a - x_b)x' + (y_a - y_b)y'] \right\} dx'\, dy'}{\iint_{-\infty}^{\infty} S(x', y')\, dx'\, dy'}$$

$$= \frac{\mathcal{F}\{S(x,y)\}_{k_x = \frac{k_0(x_a - x_b)}{d},\, k_y = \frac{k_0(y_a - y_b)}{d}}}{\iint_{-\infty}^{\infty} S(x', y')\, dx'\, dy'}, \tag{2.81}$$

where (x_a, y_a) is the location of point A.

Example 2.3: Double-pinhole interference

Consider a double-pinhole interference experiment. The geometry is shown in Fig. 2.13. Two pinholes A and B are located at $x_a = 1$ mm, $y_a = 0$ and $x_b = -1$ mm, $y_b = 0$, respectively. The light source is a lamp of extended size with a good bandpass filter centered at 600 nm. We can improve the spatial coherence by placing a small circular aperture against the light source. The distance between the aperture and the two pinholes is $d = 20$ cm. We shall calculate the largest available aperture diameter so that the visibility of the resulting interference fringe is larger than 0.5.

The light source with the circular aperture is given by $S(x, y) = \text{circ}(r/r_0)$, where $r = \sqrt{x^2 + y^2}$. According to Eq. (2.81), the complex degree of coherence is calculated as

$$\gamma_{AB} \propto \text{Fourier transform of circ}(r/r_0) = 2\frac{J_1(r_0 k_r)}{r_0 k_r}, \tag{2.82}$$

where $k_r = (k_0/d)\sqrt{(x_a - x_b)^2 + (y_a - y_b)^2}$, and $J_1(\cdot)$ is a Bessel function of the first kind, order 1. So γ_{AB} is a function of r_0 and is plotted in Fig. 2.14. The corresponding MATLAB code is listed in Table 2.2. It is shown in Fig. 2.14 that r_0

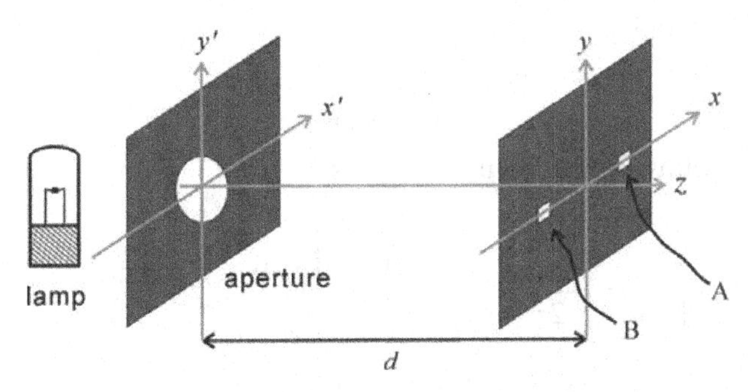

Figure 2.13 Double-pinhole experiment to evaluate coherence.

Table 2.2 *MATLAB code for plotting the modulus of the complex degree of coherence, see Fig. 2.14*

```
close all; clear all;
x=2; % xa-xb=2 mm
d=200; % d= 200 mm
l=0.6*10^(-3);          %center wavelength 600nm
r0=0.001:0.0005:0.1; %radius of the pinhole aperture
z=2*pi*x.*r0/d/l;
gama=2*besselj(1,z)./z;
gama=abs(gama./max(gama));
plot(r0,gama)
title('Complex degree of coherence')
xlabel('Radius (mm)')
ylabel('|\gamma_A_B|')
```

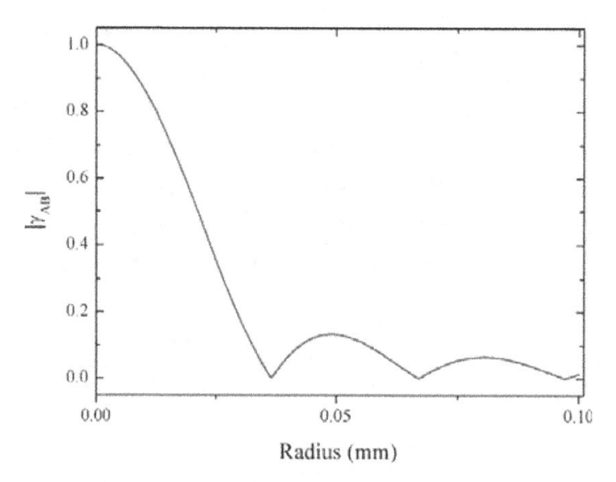

Figure 2.14 Plot of the modulus of the complex degree of coherence as a function of the radius of the circular aperture, r_0.

must be smaller than 0.021 mm (i.e., diameter 42 μm) so that the fringe visibility can be larger than 0.5.

Problems

2.1 Verify the following convolution results,

$$\exp\left[\frac{jk_0}{2z_0}(x^2+y^2)\right] * \exp\left[\frac{-jk_0}{2z}(x^2+y^2)\right] = \begin{cases} \exp\left[\frac{jk_0}{2(z_0-z)}(x^2+y^2)\right] & \text{for } z \neq z_0 \\ \delta(x,y) & \text{for } z = z_0, \end{cases}$$

which appear in Eq. (2.7).

2.2 With reference to Fig. 2.5 and Fig. 2.6 for the recording and reconstruction geometry of point sources, let $a = b = 0$, i.e., the reference and reconstruction point sources are on the z-axis, and show that, for $R \gg d$, the lateral magnification for the virtual image is [6]

$$M_{Lat}^{v} = \frac{l_1 l_2}{l_1 l_2 + R l_1 - R l_2},$$

and the longitudinal magnification is

$$M_{Long}^{v} = \frac{(l_1 l_2)^2}{(l_1 l_2 + R l_1 - R l_2)^2}.$$

2.3 Show that for a quasi-monochromatic light oscillating at ω_0 of finite size wave trains with initial phase distributed randomly between 0 and 2π within some fixed time, i.e., the phase randomly changes every time interval τ_0 and remains stable between the changes, as shown in Fig. 2.9, the complex degree of coherence is given by

$$\gamma(\tau) = \Lambda\left(\frac{\tau}{\tau_0}\right) e^{j\omega_0 \tau}.$$

2.4 According to the definition of coherence time given by Mandel, show that the coherence time of the quasi-monochromatic light from Problem 2.3 is given by $\tau_c = 2\tau_0/3$.

2.5 The typical bandwidth of a commercial He–He laser operated at $\lambda_0 = 632.8$ nm is about 500 MHz. Calculate the corresponding coherence length.

2.6 A bandpass filter is usually applied in association with a broadband light source to produce interference. Typically, the full width half-maximum (FWHM) of the transmission band is 10 nm while the center wavelength of the band is $\lambda_0 = 630$ nm. Calculate the corresponding coherence length.

2.7 When we investigate the interference formed by two incoherent light sources shown in Fig. 2.11, τ_1 and τ_2 are the time delays between the waves at point A and point B, respectively from source 1 and source 2. Show that $\tau_2 - \tau_1 = sl/dc$, assuming the small-angle approximation.

2.8 Show that for two point sources that are quasi-monochromatic, the complex degree of mutual coherence

$$|\gamma_{AB}|^2 \approx \frac{1}{2}[1 + \cos\, \omega(\tau_1 - \tau_2)]\Lambda\left(\frac{\tau_1}{\tau_0}\right)\Lambda\left(\frac{\tau_2}{\tau_0}\right)$$

if

$$\tau_1 \approx \tau_2 \gg |\tau_1 - \tau_2|.$$

2.9 Start from the Fresnel diffraction formula [Eq. (1.35)] and assume a spherical wave passing through an aperture $S(x', y')$ converging to (x_b, y_b) at the diffraction plane [Fig. 2.12(b)]. Find the diffracted field as given by the result in Eq. (2.80).

2.10 Show that the complex degree of coherence of the light produced by a uniformly quasi-monochromatic incoherent source with a shape of circular disk of radius r_0 is proportional to $J_1(\cdot)/(\cdot)$ given by Eq. (2.82). Note that since the problem is of circular symmetry, it is convenient to express the Fourier transform as follows.

$$\mathcal{F}\{f(x,y)\} = F(k_x, k_y) = 2\pi \int_0^\infty r f(r) J_0(k_r r) dr,$$

where $r = \sqrt{x^2 + y^2}$, $k_r = \sqrt{k_x{}^2 + k_y{}^2}$, and $J_0(\cdot)$ is the zeroth-order Bessel function. The above integral is also referred to as the Fourier–Bessel transform denoted by

$$\mathcal{B}\{f(r)\} = 2\pi \int_0^\infty r f(r) J_0(k_r r) dr.$$

References

1. D. Garbor, A new microscopic principle, *Nature* **161**, 777–778 (1948).
2. M. P. Givens, Introduction to holography, *American Journal of Physics* **35**, 1056–1064 (1967).
3. T.-C. Poon, On the fundamentals of optical scanning holography, *American Journal of Physics* **76**, 738–745 (2008).
4. J. W. Goodman, *Statistical Optics* (John Wiley & Sons, New York, 1985).
5. L. Mandel, Fluctuations of photon beams: the distribution of the photo-electrons, *Proceedings of the Physical Society* **74**, 233 (1959).
6. F. T. S. Yu, *Optical Information Processing* (John Wiley & Sons, New York, 1983).

3
Types of holograms

In this chapter we will introduce some basic types of holograms and their basic principles. Some topics, such as holographic recording materials, are not relevant to digital holography and hence are not covered here (for reviews of holographic materials, see Refs. [1–4]).

3.1 Gabor hologram and on-axis (in-line) holography

The recording setup of a Gabor hologram [5] is illustrated in Fig. 3.1(a). The setup is simple: we use a single beam to illuminate the object and behind the object

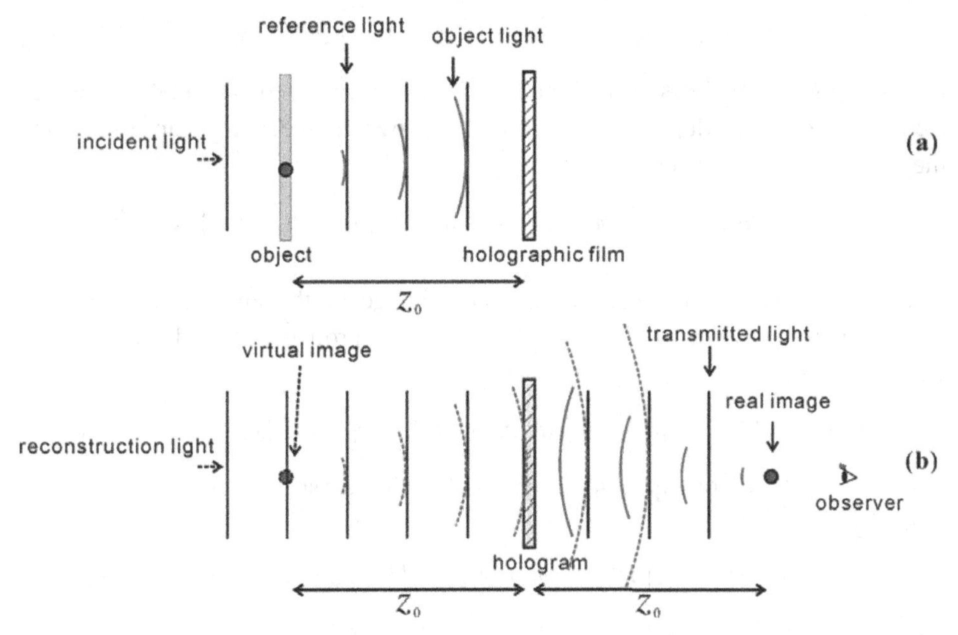

Figure 3.1 (a) Recording geometry, and (b) reconstruction geometry for the Gabor hologram.

a photosensitive material or device is placed as a recording medium. The method is effective only when the object is limited to a small fluctuation $\Delta(x, y)$ of a uniform transparency σ_0 so that the object wave scattered by the object does not disturb the uniform reference wave excessively. Hence the total transparency of the object $\sigma(x, y)$ can be expressed as

$$\sigma(x, y) = \sigma_0 + \Delta(x, y) \tag{3.1}$$

with the condition that $\Delta(x, y) \ll \sigma_0$. Therefore, the complex field on the hologram plane, at a distance z_0 away from the object, can be expressed as

$$u(x, y) = A[\sigma_0 + \Delta(x, y)] * h(x, y; z_0) = \psi_c + \psi_0(x, y), \tag{3.2}$$

where A is the amplitude of the incident light, and $h(x, y; z_0)$ is the spatial impulse response [Eq. (1.34)]. ψ_c is a uniform transmitting field, which can serve as the reference light and the scattered field, $\psi_0(x, y)$, as the amplitude fluctuation is regarded as the object field. As a result, the intensity on the hologram plane or the hologram with amplitude transmittance can be expressed as

$$t(x, y) = |\psi_c + \psi_0(x, y)|^2 = |\psi_c|^2 + |\psi_0(x, y)|^2 + \psi_c\psi_0^*(x, y) + \psi_c^*\psi_0(x, y). \tag{3.3}$$

In the reconstruction process, the hologram, $t(x, y)$, is illuminated by a plane wave of the reconstruction light. The complex field emerging from the hologram is then proportional to

$$|\psi_c|^2 + |\psi_0(x, y)|^2 + \psi_c\psi_0^*(x, y) + \psi_c^*\psi_0(x, y). \tag{3.4}$$

According to the analysis in Section 2.2, $\psi_c\psi_0^*(x, y)$ reconstructs a real image, i.e., a complex conjugate duplicate of the object (the twin image) at a distance z_0 behind the hologram. We can see it clearly as

$$\psi_c\psi_0^*(x, y) * h(x, y; z_0) = \psi_c[A\Delta(x, y) * h(x, y; z_0)]^* * h(x, y; z_0) \propto \Delta^*(x, y).$$

Similarly, $\psi_c^*\psi_0(x, y)$ reconstructs a virtual image of the amplitude fluctuation $\Delta(x, y)$ at the location of the object. After back-propagating a distance of z_0, we have

$$\psi_c^*\psi_0(x, y) * h^*(x, y; z_0) = \psi_c^*[A\Delta(x, y) * h(x, y; z_0)] * h^*(x, y; z_0) \propto \Delta(x, y).$$

The first two terms of Eq. (3.4) correspond to the transmitted zeroth-order beam given by

$$\left[|\psi_c|^2 + |\psi_0(x, y)|^2\right] * h(x, y; z),$$

where z is any distance behind the hologram. The reconstruction is shown in Fig. 3.1(b).

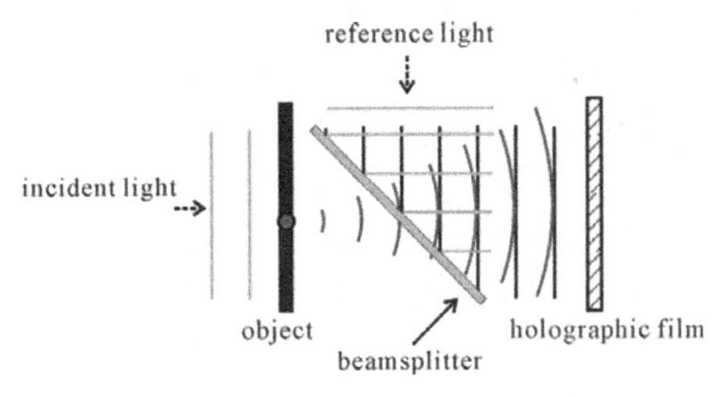

Figure 3.2 Recording geometry for on-axis holography.

The merit of Gabor holography is that the setup is very simple and it is possible to produce a Gabor hologram using a low-coherence light source. On the other hand, because all the light fields propagate along the same direction in the reconstruction stage, they are observed simultaneously, and the reconstructed image is always blurred by the transmitted zeroth-order light and the twin image. Another problem of the Gabor hologram is that the amplitude fluctuation of the object must be small enough, i.e., $\Delta(x, y) \ll \sigma_0$, to make the technique useful. Accordingly, Gabor holography cannot be applied to the usual diffusely reflecting objects. However, this shortcoming can be overcome by using an independent reference light as a reference, as shown in Fig. 3.2. Since the reference light and the object light overlap along the same direction, this setup is called the on-axis or in-line geometry. In on-axis holography, there is no limitation on the types of the object used. However, the problem of the zeroth-order light and the twin image remains.

3.2 Off-axis holography

The recording setup of off-axis holography [6] is illustrated in Fig. 3.3(a). In the geometry, the reference light is a plane wave with an offset angle, θ, with respect to the recording film. Assuming that the propagation vector of the reference light is on the x–z plane, the *off-axis hologram* can be expressed as

$$\begin{aligned} t(x, y) &= |\psi_0(x, y) + \psi_r e^{jk_0 \sin\theta x}|^2 \\ &= |\psi_0(x, y)|^2 + |\psi_r|^2 + \psi_0(x, y)\psi_r^* e^{-jk_0 \sin\theta x} + \psi_0^*(x, y)\psi_r e^{jk_0 \sin\theta x} \end{aligned} \tag{3.5a}$$

or

$$t(x, y) = |\psi_0(x, y)|^2 + |\psi_r|^2 + 2|\psi_0^*(x, y)\psi_r| \cos\left[2\pi f_x x + \phi(x, y)\right], \tag{3.5b}$$

where $\psi_0(x, y)$ and $\psi_r e^{jk_0 \sin\theta x}$ are the complex amplitudes of the object light and the reference light, respectively, on the holographic film. $\phi(x, y)$ is the phase angle of $\psi_0^*(x, y)\psi_r$ as $\psi_0^*(x, y)\psi_r = |\psi_0^*(x, y)\psi_r| e^{j\phi(x, y)}$. Finally, $f_x = \sin\theta / \lambda_0$ is called the *spatial carrier frequency* of the hologram, reminiscent of the terminology used in

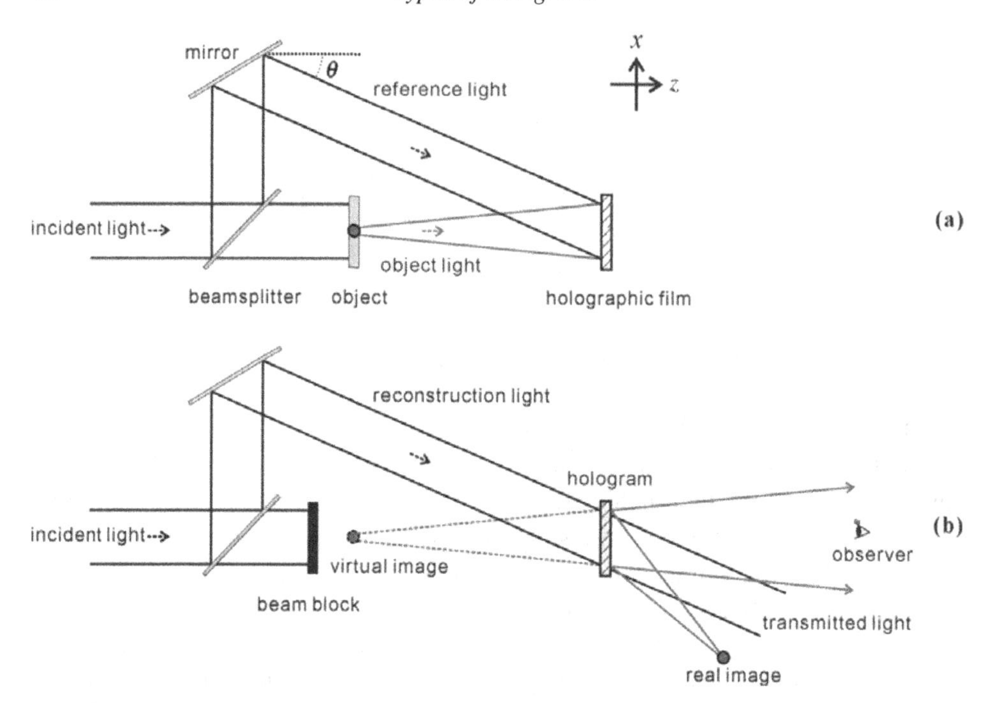

Figure 3.3 (a) Recording setup, and (b) reconstruction setup for off-axis holography.

communication systems where a carrier wave is used to "carry" the message in the theory of modulation. Hence the off-axis hologram is also called the *carrier-frequency hologram.*

In the reconstruction, the hologram is illuminated by the reconstruction light along the same direction as the reference light, as shown in Fig. 3.3(b). Assuming the amplitude of the reconstruction light is the same as that of the reference light, the complex field just behind the hologram can be expressed as

$$\psi_r e^{jk_0 \sin\theta x} \times t(x,y) = |\psi_r|^2 \psi_r e^{jk_0 \sin\theta x} + |\psi_0(x,y)|^2 \psi_r e^{jk_0 \sin\theta x} + \psi_0(x,y)|\psi_r|^2 + \psi_0^*(x,y)\psi_r^2 e^{j2k_0 \sin\theta x}, \tag{3.6}$$

where we have used Eq. (3.5a) to write the above equation. Similar to Gabor holography, the first two terms on the right hand side of Eq. (3.6) represent the transmitted light, i.e., the zeroth-order beam. The third term contributes to the virtual reconstructed image. Finally, the last term contributes to the conjugate real image. To understand how the virtual image can be viewed without the annoying disturbance due to the transmitted light and the twin image, it is convenient to analyze the situation in the Fourier spectrum domain. In the spectrum domain, the four terms on the right side of Eq. (3.6) can be respectively calculated by

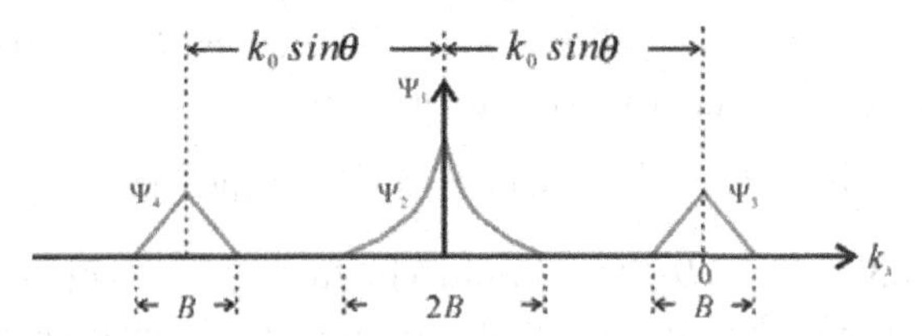

Figure 3.4 One-dimensional plot of the spectrum of the off-axis hologram. Note that Ψ_3, the spectrum of the virtual image, is centered at $k_x = 0$.

$$
\begin{aligned}
\Psi_1(k_x, k_y) &= \mathcal{F}\{|\psi_r|^2 \psi_r e^{jk_0 \sin\theta x}\} = \mathcal{F}\{|\psi_r|^2 \psi_r\} * \delta(k_x + k_0 \sin\theta, k_y),\\
\Psi_2(k_x, k_y) &= \mathcal{F}\{|\psi_0(x, y)|^2 \psi_r\} * \delta(k_x + k_0 \sin\theta, k_y),\\
\Psi_3(k_x, k_y) &= \mathcal{F}\{\psi_0(x, y)|\psi_r|^2\},\\
\Psi_4(k_x, k_y) &= \mathcal{F}\{\psi_0^*(x, y)\psi_r^2\} * \delta(k_x + 2k_0 \sin\theta, k_y).
\end{aligned}
$$

Since ψ_r is a constant, Ψ_1 represents a pulse at $k_x = -k_0\sin\theta$, $k_y = 0$; Ψ_2 is proportional to the auto-correlation of $\mathcal{F}\{\psi_0(x, y)\}$, also located at $k_x = -k_0\sin\theta$, $k_y = 0$. The third term Ψ_3 is proportional to $\mathcal{F}\{\psi_0(x, y)\}$ and is at $k_x = 0$, $k_y = 0$. Finally, the last term Ψ_4 is proportional to $\mathcal{F}\{\psi_0^*(x, y)\}$ and is at $k_x = -2k_0\sin\theta$, $k_y = 0$. We plot a one-dimensional version of the four spectra in Fig. 3.4. In the figure the bandwidths of Ψ_3 and Ψ_4 are considered to be B. Thus the bandwidth of Ψ_2 is $2B$. Apparently, the spectrum of the virtual image can be isolated, provided the offset angle θ is large enough. As a result, we can observe a clear virtual image because the directions of the transmitted light and the twin image deviate from the optical axis, i.e., along the z-axis, as shown in Fig. 3.3(b).

The minimum offset angle, θ_{min}, can be determined from inspection of Fig. 3.4. To avoid overlapping of spectra Ψ_2 and Ψ_3, we have to let

$$
k_0 \sin\theta \geq \frac{3}{2}B
$$

Accordingly, the minimum offset angle is determined to be

$$
\theta_{min} = \sin^{-1}\left(\frac{3B}{2k_0}\right). \tag{3.7}
$$

Example 3.1: Determination of the offset angle and the required resolving power of the recording medium

In this example, we consider the recording of an off-axis hologram using a He–Ne laser ($\lambda_0 = 0.6328$ μm) and the viewing angle of the reconstructed virtual image must be at least 20°. What is the minimum offset angle, and what is the minimum resolving power of the recording medium required?

The *viewing angle* is the maximum angle at which the reconstructed image can be viewed (we will discuss this further in Chapter 7). Since the required largest angle of the reconstructed image is 20° / 2 = 10°, the maximum spread of the propagation vector along the x-direction is then $k_{0x} = k_0 \times \sin(10°)$. Thus the bandwidth of the reconstructed virtual image is

$$B = 2 \times k_0 \times \sin(10°) = 3448 \text{ rad/mm},$$

where $k_0 = 2\pi/0.6328\ \mu\text{m} = 9929$ rad/mm. The factor of two in the above equation corresponds to taking into account that the spread of the light from the reconstructed image also spills into the negative x-direction. By substituting the calculated value of B into Eq. (3.7), we can determine θ_{min} as

$$\theta_{min} = \sin^{-1}\left(\frac{3 \times 3448 \text{ rad/mm}}{2 \times 9929 \text{ rad/mm}}\right) = 31.4°.$$

Now, in order to successfully record the off-axis hologram, the recording medium must be able to resolve the spatial carrier frequency, $f_x = \sin\theta_{min}/\lambda_0$, plus half the bandwidth, $B/2$, of the reconstructed image, i.e., the required resolvable spatial frequency is

$$f_{resolvable} = \frac{\sin\theta_{min}}{\lambda_0} + \frac{B/2}{2\pi} = \frac{\sin\theta_{\text{min}}}{\lambda_0} + \frac{B}{4\pi}. \tag{3.8}$$

With the calculated values of θ_{min} and B, we can find

$$\begin{aligned} f_{resolvable} &= \frac{\sin 31.4°}{0.6328\ \mu\text{m}} + \frac{3448 \text{ rad/mm}}{2\pi} \\ &= 1095 \text{ cycle/mm or line–pair/mm or lp/mm.} \end{aligned}$$

This resolution can be achieved by most available holographic emulsions, typically several thousand line-pairs per millimeter. However, such resolution is well beyond the capability of existing electronic photosensitive devices such as CCDs or spatial light modulators (SLMs). We will discuss SLMs and holographic three-dimensional display in Chapter 7.

3.3 Image hologram

An *image hologram* can be recorded as shown in Fig. 3.5(a), where a real image of the object is formed on the holographic film using a lens. The light emerging from the real image serves as the object wave. Mathematically, on recording, the image hologram is given by

$$t(x, y) = |\psi_i(x, y) + \psi_r e^{jk_0 \sin\theta x}|^2, \tag{3.9}$$

where $\psi_i(x, y)$ represents the complex field of the real image on the hologram and $\psi_r e^{jk_0 \sin\theta x}$ is the off-axis reference plane wave. Upon reconstruction and assuming the amplitude of the reconstruction light is the same as that of the reference light as shown in Fig. 3.5(b), the complex field just behind the hologram can be expressed as

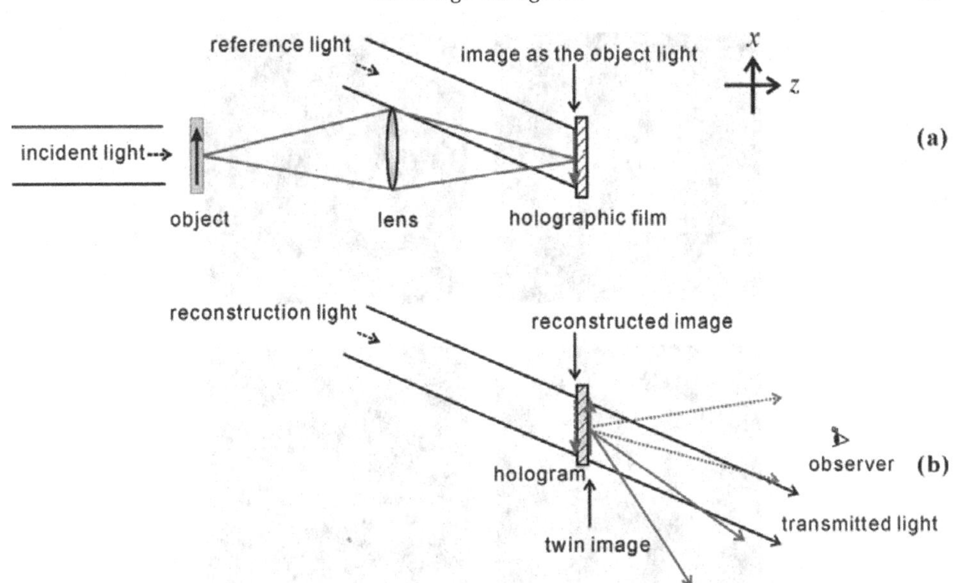

Figure 3.5 (a) Recording geometry, and (b) reconstruction geometry for the image hologram.

$$t(x,y) \times \psi_r e^{jk_0 \sin\theta x} = [|\psi_i|^2 + |\psi_r|^2]\psi_r e^{jk_0 \sin\theta x} + \psi_i(x,y)|\psi_r|^2 + \psi_i^*(x,y)\psi_r^2 e^{j2k_0 \sin\theta x}. \quad (3.10)$$

The first term of the right hand side of the above equation makes up the zeroth-order beam propagating along the direction of the reconstruction beam. The second term is the reconstructed image and is on the hologram plane. The third term is the twin image, which is reconstructed on the hologram plane as well but propagating along the direction that is 2θ away from the z-axis, but the observer views the reconstruction along the z-axis.

We cannot observe the reconstructed image of a conventional off-axis hologram using a polychromatic light source, such as a lamp, because there is a serious chromatic aberration in the hologram. As we have shown in Section 2.3.3, chromatic aberration is proportional to the distance between the object and the hologram plane. If the object is reconstructed on the plane of the hologram, chromatic aberration can be minimized and the reconstructed images from various wavelengths overlap, producing a clear, white reconstructed image. Therefore, the image hologram can be reconstructed using either monochromatic light or polychromatic light. It should be noted that the twin image is also reconstructed on the hologram plane, as shown in Fig. 3.5(b). Thus the off-axis geometry is necessary for recording the image hologram and only the reconstructed image can be observed along the direction of the observer.

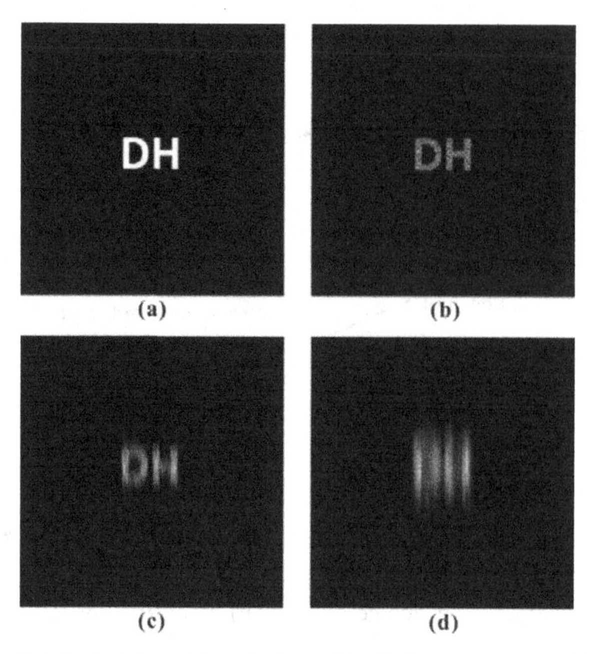

Figure 3.6 Original object (a) and the white-light reconstructed images with distances between the image and hologram of (b)1 cm, (c) 5 cm, and (d)15 cm, respectively.

Example 3.2: Simulation of an image hologram

In this example, we would like to demonstrate the reconstruction of an image hologram using white light. To simplify the simulation, we only consider a single diffraction order (the virtual image) of an off-axis hologram. Thus the off-axis zeroth-order light and the twin image are ignored in the simulation. The object pattern shown in Fig. 3.6(a) contains 256 × 256 pixels. The pixel pitch is 50 μm, and the recording wavelength is 0.65 μm. The offset angle is 10°. Here we do not discuss the details of the discrete Fourier transform and the simulation of diffraction, which will be discussed in Chapter 4.

We first obtain the diffracted field at a desired reconstruction plane using wavelengths ranging from 0.45 μm to 0.65 μm with wavelength separation of 0.005 μm. Thus we have a total of 41 reconstructed images. We then superimpose the reconstructed images of all the wavelengths, producing a multi-wavelength reconstructed image. Figure 3.6(b)–(d) are the multi-wavelength reconstructed images while the distances between the images and the hologram are 1 cm, 5 cm, and 15 cm, respectively. As the distance between the reconstructed image and the hologram is close (i.e., 1 cm), the hologram can be regarded as an image hologram and we can see that the reconstructed image is sharp even though the reconstruction light is white light. However, as the reconstructed image is far from the hologram plane,

Table 3.1 *MATLAB code for simulation of an image hologram, see Example 3.2*

```
% Input data, set parameters
clear all, close all;
Ii=imread('DH256.bmp');%256×256 pixels 8bit
figure; imshow(Ii);
title('Object pattern')
axis off
Ii=double(Ii);
PH=rand([256,256]);
Ii=Ii.*exp(2i*pi*PH); %add a random phase on the object
M=512;
I=zeros(512);
I(128:383,128:383)=Ii; %zero padding
z=15; %object distance in cm
w=6500*10^-8; %(cm, wavelength)
delta=0.005;  %cm, pixel size 50um
r=1:M;
c=1:M;
[C, R]=meshgrid(c, r);

% Forward propagation (650nm)
p=exp(-2i*pi*z.*((1/w)^2-(1/M/delta)^2.*(C-M/2-1).^2-...
    (1/M/delta)^2.*(R-M/2-1).^2).^0.5);
A0=fftshift(ifft2(fftshift(I)));
Az=A0.*p;
E=fftshift(fft2(fftshift(Az))); %1st order of the hologram

% Reconstruction (650nm)
p=exp(-2i*pi*(-z).*((1/w)^2-(1/M/delta)^2.*...
    (C-M/2-1).^2-(1/M/delta)^2.*(R-M/2-1).^2).^0.5);
A1=fftshift(ifft2(fftshift(E)));
Az1=A1.*p;
R1=fftshift(fft2(fftshift(Az1)));
R1=(abs(R1)).^2;
figure; imshow(R1/max(max(R1)));
title('Reconstructed image(650nm)')
axis off

% Reconstruction (450nm~650nm)
dw=50;
IMA=zeros(512,512);
for g=0:40;
    w2=(6500-dw*g)*10^-8; %reconstruction wavelength
    E2=E.*exp(2i*pi*sind(10)*(w-w2)/w/w2.*R*delta);
    % phase mismatch due to the wavelength shift
    p=exp(-2i*pi*(-z).*((1/w2)^2-...
    (1/M/delta)^2.*(C-M/2-1).^2-...
    (1/M/delta)^2.*(R-M/2-1).^2).^0.5);
    Az2=ifft2(fftshift(E2)).*(fftshift(p));
```

Table 3.1 (*cont.*)

```
    R2=fftshift(fft2(Az2));
    R2=(abs(R2)).^2;   %summation of all wavelengths
    IMA=IMA+R2;
end
IMA=IMA/max(max(IMA));
figure; imshow(IMA)
title('Reconstructed image(white light)')
axis off
```

chromatic aberration becomes large and the reconstructed image is blurred, as shown in Figs. 3.6(c) and (d). The MATLAB code is listed in Table 3.1 as a reference.

3.4 Fresnel and Fourier holograms

3.4.1 Fresnel hologram and Fourier hologram

When the object wave on the hologram plane is described by Fresnel diffraction of the object [Eq. (1.35)], we have a *Fresnel hologram*. On the other hand, the hologram is called a *Fourier hologram* if the object wave on the hologram plane is described by the Fourier transform of the object. The Fourier hologram can be recorded using the setup shown in Fig. 3.7(a). In the setup the object is set at the front focal plane of lens 2. Thus the optical field at the back focal plane of lens 2 is the Fourier transform of the object field distribution. Meanwhile, a focused light spot beside the object, through the use of lens 1, is used as a reference light, making a tilted plane wave on the holographic film.

According to Eq. (1.45), the total optical field at the back focal plane of lens 2 can be expressed as

$$\psi_t(x,y) = \mathcal{F}\left\{\sigma_0(x,y) + A\delta(x-x_0,y)\right\}_{\substack{k_x = k_0x/f \\ k_y = k_0y/f}}$$
$$= \Sigma_0\left(\frac{k_0x}{f},\frac{k_0y}{f}\right) + A\exp\left(\frac{jk_0x_0x}{f}\right), \tag{3.11}$$

where $\sigma_0(x, y)$ is the amplitude transmittance of the object and Σ_0 (k_x, k_y) is the Fourier transform of $\sigma_0(x, y)$; $\delta(x - x_0, y)$ stands for the reference light spot at $x = x_0$, $y = 0$; f is the focal length of lens 2, and A is the amplitude of the reference light. Consequently, the hologram can be expressed as

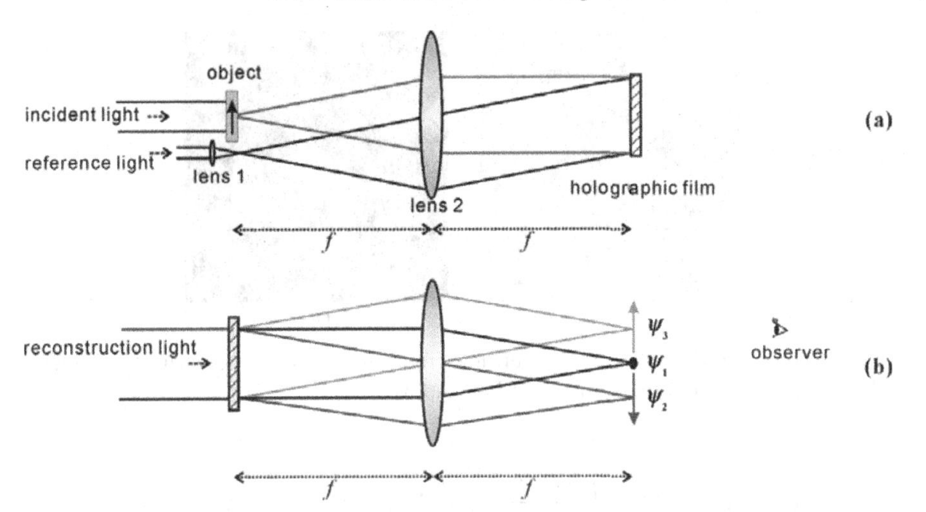

Figure 3.7 (a) Recording geometry, and (b) reconstruction geometry for a Fourier hologram.

$$
\begin{aligned}
t(x,y) = {} & \left|\Sigma_0\left(\frac{k_0x}{f},\frac{k_0y}{f}\right)\right|^2 + |A|^2 + \Sigma_0\left(\frac{k_0x}{f},\frac{k_0y}{f}\right)\times A^*\exp\left(\frac{-jk_0x_0x}{f}\right) \\
& + \Sigma^*\left(\frac{k_0x}{f},\frac{k_0y}{f}\right)\times A\exp\left(\frac{jk_0x_0x}{f}\right).
\end{aligned} \tag{3.12}
$$

In the reconstruction process, the hologram placed at the front focal plane of a lens is illuminated using a normal-incident plane wave with unit amplitude, as shown in Fig. 3.7(b). According to Eq. (3.12), the complex field at the back focal plane of the lens contains three terms:

$$
\begin{aligned}
\psi_1(x,y) & = \mathcal{F}\left\{\left|\Sigma_0\left(\frac{k_0x}{f},\frac{k_0y}{f}\right)\right|^2 + |A|^2\right\}\Bigg|_{\substack{k_x = k_0x/f \\ k_y = k_0y/f}} \\
& = \frac{f^4}{k_0^4}\sigma_0(-x,-y)\otimes\sigma_0(-x,-y) + \frac{4\pi^2f^2}{k_0^2}|A|^2\delta(x,y),
\end{aligned} \tag{3.13a}
$$

$$
\begin{aligned}
\psi_2(x,y) & = \mathcal{F}\left\{\Sigma_0\left(\frac{k_0x}{f},\frac{k_0y}{f}\right)\times A^*\exp\left(\frac{-jk_0x_0x}{f}\right)\right\}\Bigg|_{\substack{k_x = k_0x/f \\ k_y = k_0y/f}} \\
& = \frac{f^2}{k_0^2}A^*\sigma_0(-x+x_0,-y),
\end{aligned} \tag{3.13b}
$$

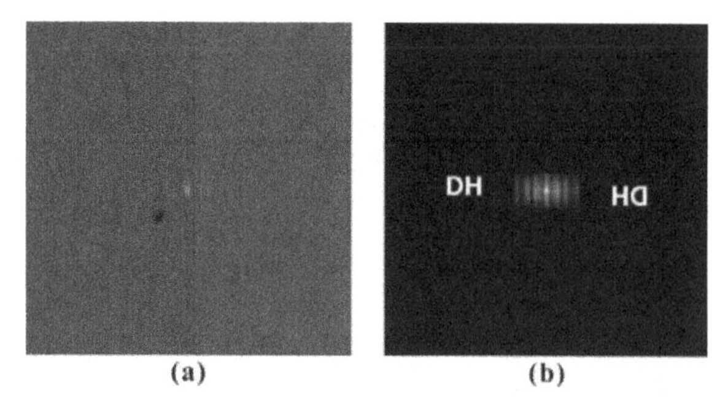

Figure 3.8 (a) The off-axis Fourier hologram, and (b) the reconstructed image. The object pattern is the same as in Fig. 3.6(a).

$$\psi_3(x,y) = \mathcal{F}\left\{\Sigma_0^*\left(\frac{k_0x}{f},\frac{k_0y}{f}\right)\times A\exp\left(\frac{jk_0x_0x}{f}\right)\right\}\Bigg|_{\substack{k_x = k_0x/f\\ k_y = k_0y/f}}$$

$$= \frac{f^2}{k_0^2}A\sigma_0^*(x+x_0,y). \tag{3.13c}$$

ψ_1 is the zeroth-order beam located at $x = 0$, $y = 0$ on the back focal plane. ψ_2 is the inverted reconstructed image at $x = x_0$, $y = 0$, and ψ_3 is the conjugate image at $x = -x_0$, $y = 0$. So ψ_1, ψ_2, and ψ_3 can be separated from each other, as shown in Fig. 3.7(b).

Example 3.3: Simulation of a Fourier hologram

In this example we would like to simulate the recording and reconstruction of a Fourier hologram. First the complex field of the object pattern on the focal plane can be obtained by performing a Fourier transform. Then, a tilted plane-wave reference light is added to the complex field, producing a Fourier hologram as shown in Fig. 3.8(a). The Fourier hologram can be reconstructed by applying a Fourier transform. The reconstructed image is shown in Fig. 3.8(b). It can be seen that both the reconstructed image and the twin image are clear on the reconstruction plane. The separation between the three diffraction orders on the reconstruction plane depends on the offset angle of the reference light. Too small an offset angle will result in crosstalk between adjacent diffraction orders. The MATLAB code is listed in Table 3.2.

3.4.2 Lensless Fourier hologram

A Fourier hologram can also be recorded using the geometry shown in Fig. 3.9. Because the Fourier transforming lens is not involved in the geometry, the hologram thus obtained is called a *lensless Fourier hologram*. In the geometry,

Table 3.2 *MATLAB code for simulation of a Fourier hologram, see Example 3.3*

```
% Input data, set parameters
clear all, close all;
Ii=imread('DH256.bmp');%256×256 pixels,8bit image
Ii=double(Ii);
M=512;
I=zeros(512);
I(128:383,128:383)=Ii; % zero-padding
figure; imshow(mat2gray(abs(I)));
title('Object pattern')
axis off

% Produce the Fourier hologram
r=1:M;
c=1:M;
[C, R]=meshgrid(c, r);
O=fftshift(ifft2(fftshift(I)));
R=ones(512,512);
R=R*max(max(abs(O)));
R=R.*exp(2i*pi.*C/4); % tilted reference light
H=(abs(O+R)).^2; % Fourier hologram
figure; imshow(mat2gray(abs(H)));
title('Fourier hologram')
axis off

% Reconstruction
U=fftshift(ifft2(fftshift(H)));
figure; imshow(900.*mat2gray(abs(U)));
title('Reconstructed image')
axis off
```

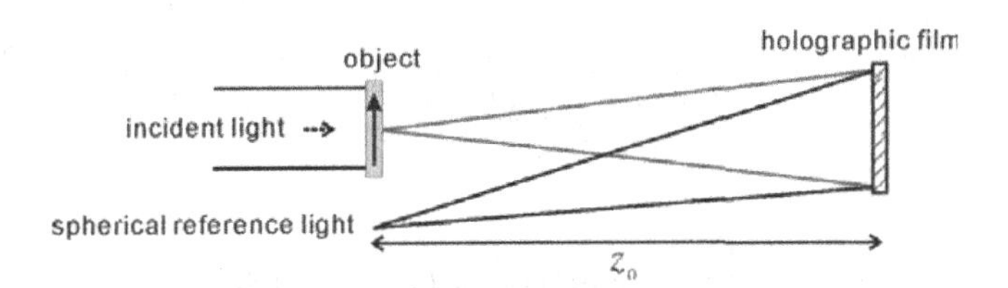

Figure 3.9 Recording geometry for a lensless Fourier hologram.

both the object and a focused light spot are still located on the same plane. Suppose that the distance between the object, $\sigma_0(x, y)$, and the holographic film is within the Fresnel region, so the complex amplitudes of the object light $\psi_0(x, y)$ and the reference light $\psi_r(x, y)$ at the hologram plane can be described using the Fresnel diffraction formula. Explicitly, they are expressed as

$$\psi_0(x,y) = \exp\left[\frac{-jk_0}{2z_0}(x^2+y^2)\right] \times \mathcal{F}\left\{\sigma_0(x,y)\exp\left[\frac{-jk_0}{2z_0}(x^2+y^2)\right]\right\}_{k_x=\frac{k_0x}{z_0},\ k_y=\frac{k_0y}{z_0}} \tag{3.14a}$$

$$\psi_r(x,y) = \exp\left[\frac{-jk_0}{2z_0}(x^2+y^2)\right] \times \mathcal{F}\left\{\delta(x-x_0,y)\exp\left[\frac{-jk_0}{2z_0}(x^2+y^2)\right]\right\}_{k_x=\frac{k_0x}{z_0},\ k_y=\frac{k_0y}{z_0}}$$

$$= \exp\left\{\frac{-jk_0}{2z_0}\left[(x-x_0)^2+y^2\right]\right\}, \tag{3.14b}$$

where z_0 is the distance between the object and the hologram. Note that the proportional constants in ψ_0 and ψ_r have been dropped for simplicity. The recorded hologram becomes

$$t(x,y) = |\psi_0(x,y) + \psi_r(x,y)|^2.$$

Our concern is the cross terms upon holographic recording, which are

$$\begin{aligned}\psi_0(x,y)\psi_r^*(x,y) + \psi_0^*(x,y)\psi_r(x,y) &= \exp\left(\frac{-jk_0x_0x}{z_0} + \frac{jk_0x_0^2}{2z_0}\right)\\ &\times \mathcal{F}\left\{\sigma_0(x,y)\exp\left[\frac{-jk_0}{2z_0}(x^2+y^2)\right]\right\}_{k_x=\frac{k_0x}{z_0},\ k_y=\frac{k_0y}{z_0}}\\ &+\exp\left(\frac{jk_0x_0x}{z_0} - \frac{jk_0x_0^2}{2z_0}\right)\\ &\times \mathcal{F}\left\{\sigma_0(x,y)\exp\left[\frac{-jk_0}{2z_0}(x^2+y^2)\right]\right\}^*_{k_x=\frac{k_0x}{z_0},\ k_y=\frac{k_0y}{z_0}}.\end{aligned} \tag{3.15}$$

We can now use the setup shown in Fig. 3.7(b) to reconstruct the lensless Fourier hologram. Ignoring the zeroth-order light, the optical field at the back focal plane of the lens is

$$\mathcal{F}\left\{\psi_0(x,y)\psi_r^*(x,y) + \psi_0^*(x,y)\psi_r(x,y)\right\}_{k_x=\frac{k_0x}{f},\,k_y=\frac{k_0y}{f}} \propto \sigma_0\left(-\frac{z_0}{f}x + x_0, -\frac{z_0}{f}y\right)$$
$$\times\exp\left\{\frac{-jk_0}{2z_0}\left[\left(\frac{z_0}{f}x-x_0\right)^2 + \left(\frac{z_0}{f}y\right)^2\right]\right\} + \sigma_0^*\left(\frac{z_0}{f}x + x_0, \frac{z_0}{f}y\right)$$
$$\times\exp\left\{\frac{jk_0}{2z_0}\left[\left(\frac{z_0}{f}x + x_0\right)^2 + \left(\frac{z_0}{f}y\right)^2\right]\right\}, \tag{3.16}$$

where we have neglected any constant factors. Again, we obtain an inverted image at $x = fx_0/z_0$, $y = 0$, and a conjugate image at $x = -fx_0/z_0$, $y = 0$. Note that there is an additional quadratic phase among the reconstructed images. So the reconstruction method only applies to the display of intensity images.

3.5 Rainbow hologram

The rainbow hologram, invented by Benton [7], is one of the easiest techniques to produce a monochromatic reconstructed image using white light reconstruction. The recording of the rainbow hologram contains two steps. First we record a conventional off-axis hologram with a reference light at angle θ with respect to the z-axis, as shown in Fig. 3.10(a). In the second step, we use reconstruction light at angle $-\theta$ to illuminate the first hologram, producing a real image of the object on axis [see Problem 3.1]. Such reconstruction light is called a *conjugate beam*. However, the real image is reconstructed using a very small aperture of the hologram as a narrow horizontal slit is placed next to the hologram. With this aperture-reduced real image, we record another hologram as shown in Fig. 3.10(b). The resulting hologram is the desired rainbow hologram. In the reconstruction process, we use a white conjugate reference beam to illuminate the rainbow hologram, i.e., being conjugate to the reference light used to record the second hologram. The slit together with the real image of the object is reconstructed. In other words, all the diffracted light converges to the location of the slit, forming a bright slit image. Because of the inherent chromatic aberration of the hologram, the slit images for different colors are separated vertically. If the observer's eyes locate the slit image, a bright, monochromatic reconstructed image can be seen. The color of the image changes gradually from red to blue as the eyes move vertically. Because the two-step recording process is complicated, a one-step recording technique for the rainbow hologram has

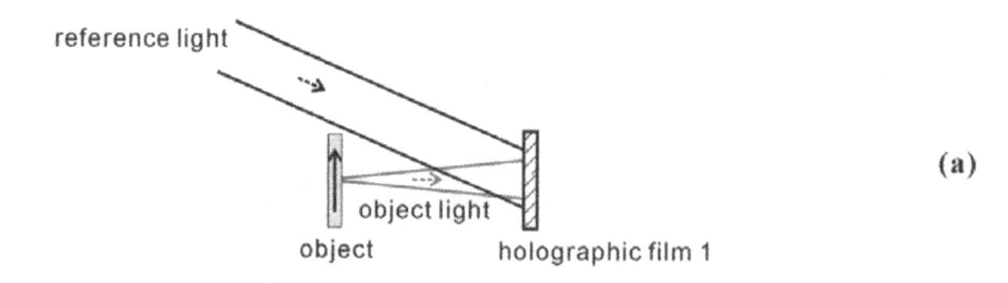

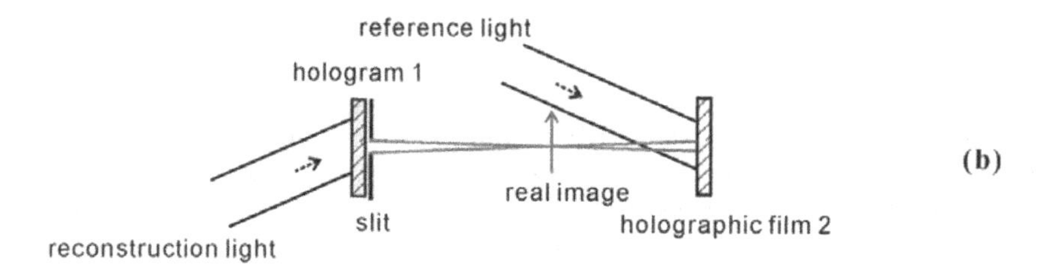

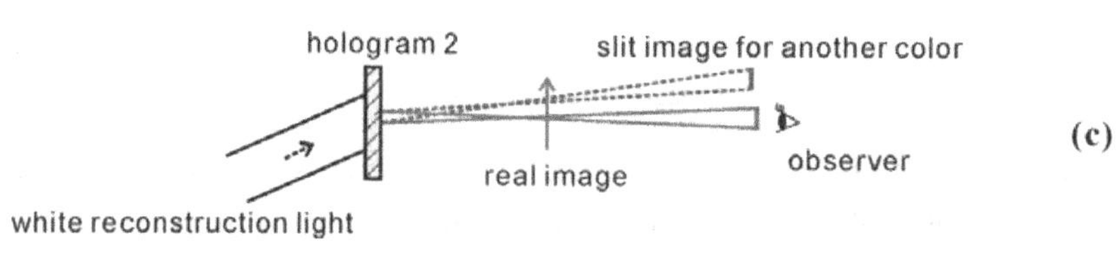

Figure 3.10 (a) First-step and (b) second-step recording geometries; (c) reconstruction geometry for the rainbow hologram.

been proposed [8, 9]. In one-step recording, the real image of the object is formed using a slit and a lens.

Example 3.4: Simulation of a rainbow hologram simulation

In this example, we would like to demonstrate the reconstruction of a rainbow hologram. To simplify the simulation, we only consider a single diffraction order (the real image) of an off-axis rainbow hologram and hence we just employ complex holograms for simulations – procedures similar to picking the relevant terms for image reconstruction, described in Chapter 2 [see Eq. (2.16)]. Thus the off-axis zeroth-order light and the twin image are ignored in the simulation. The object pattern is the same as that shown in Fig. 3.6(a). The pixel pitch is 50 μm, and the recording wavelength is 0.65 μm. The generation of a rainbow hologram includes two steps of holographic recording. We first calculate the diffracted field 30 cm away from the object as shown in Fig. 3.10(a) to obtain hologram 1. We then crop the complex field to obtain

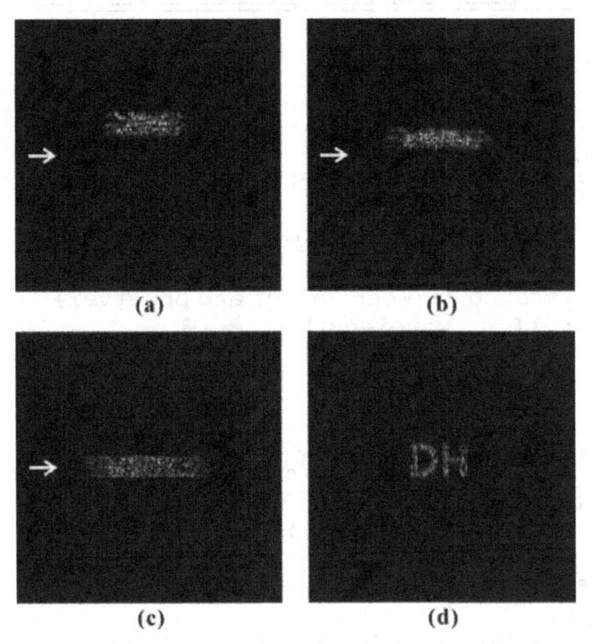

Figure 3.11 Reconstruction of the rainbow hologram. (a)–(c) The slit images for B ($\lambda_0 = 450$ nm), G ($\lambda_0 = 550$ nm), and R ($\lambda_0 = 650$ nm) colors at the observer's plane, respectively. The arrows indicate the altitude of the observer's eyes. (d) The reconstructed image at the object plane for the R color.

a slit-aperture hologram, which is illuminated by a conjugate beam to give the field propagated at a distance of 60 cm to obtain hologram 2 [see Fig. 3.10(b)]. Finally, we reconstruct hologram 2 by a conjugate beam by allowing the propagation of 30 cm to get the real image "DH" in Fig. 3.10(c) and then another 30 cm to reach the observer's pupil. In the reconstruction, we use three different colors of light, R ($\lambda_0 = 650$ nm), G ($\lambda_0 = 550$ nm), and B ($\lambda_0 = 450$ nm), to illuminate the hologram [Fig. 3.11(a–c)]. The slit images of the three colors in the observer plane are separated. Hence if the eye level of the observer moves up or down vertically, he or she will see the reconstructed image with a different color. This is also why the technique is called "rainbow holography." Thus only light with a single wavelength or with a narrow bandwidth of the reconstructed light can pass through the small pupil of the observer. Indeed Fig. 3.11(d) shows the reconstructed image of a single wavelength. Because of the mask effect of the slit, the resolution of the reconstructed image is worse than that of the conventional holographic image. This is a shortcoming of rainbow holography [10, 11]. The MATLAB code of the example is listed in Table 3.3 as a reference.

Table 3.3 *MATLAB code for simulation of a rainbow hologram, see Example 3.4*

```
% Input data, set parameters
clear all, close all;
Ii=imread('DH256.bmp');% 256×256 pixels 8bit image
Ii=double(Ii);
PH=rand([256,256]);
Ii=Ii.*exp(2i*pi*PH); %random phase on the object pattern

% 1. Forward propagation to hologram 1
M=256;
z=30; %(cm, distance between object and observer)
w=6500*10^-8; %(cm, wavelength)
delta=0.005; %(cm, pixel pitch 50um)
r=1:M;
c=1:M;
[C, R]=meshgrid(c, r);
p=exp(-2i*pi*z.*((1/w)^2-(1/M/delta)^2.*(C-M/2-1).^2-...
    (1/M/delta)^2.*(R-M/2-1).^2).^0.5);
A0=fftshift(ifft2(fftshift(Ii)));
Az=A0.*p;
E=fftshift(fft2(fftshift(Az)));
ES=zeros(512);
ES(246:265,128:383)=E(118:137,:); % A slit aperture

% 2.Forward propagation to hologram 2
M=512;
r=1:M;
c=1:M;
[C, R]=meshgrid(c, r);
A1=fftshift(ifft2(fftshift(conj(ES))));
z=60; % distance between hologram1 and hologram 2
p=exp(-2i*pi*z.*((1/w)^2-(1/M/delta)^2.*(C-M/2-1).^2-...
    (1/M/delta)^2.*(R-M/2-1).^2).^0.5);
Az1=A1.*p;
H=fftshift(fft2(fftshift(Az1)));
%object light at the hologram plane

% 3.Reconstruction (650nm)
H2=zeros(1024);
H2(256:767,256:767)=conj(H);
M=1024;
r=1:M;
c=1:M;
[C, R]=meshgrid(c, r);
z=30;
p=exp(-2i*pi*z.*((1/w)^2-(1/M/delta)^2.*(C-M/2-1).^2-...
    (1/M/delta)^2.*(R-M/2-1).^2).^0.5);
A2=fftshift(ifft2(fftshift(H2)));
Az2=A2.*p;
R650=fftshift(fft2(fftshift(Az2)));
R650=(abs(R650)).^2;
R650=R650/max(max(R650));
R650=R650(384:639,384:639);
figure; imshow(R650);
```

Table 3.3 (*cont.*)

```
title('Reconstructed image(650nm)')
axis off
Az2=Az2.*p; % at observer plane
S650=fftshift(fft2(fftshift(Az2)));
S650=(abs(S650)).^2;
S650=S650/max(max(S650));
S650=S650(384:639,384:639);

% Reconstruction (550nm)
w2=5500*10^-8;
H3=H2.*exp(2i*pi*sind(0.4)*(w-w2)/w/w2.*R*delta);
p=exp(-2i*pi*z.*((1/w2)^2-(1/M/delta)^2.*(C-M/2-1).^2-...
    (1/M/delta)^2.*(R-M/2-1).^2).^0.5);
Az3=fftshift(ifft2(fftshift(H3))).*p;
R550=fftshift(fft2(fftshift(Az3)));
R550=(abs(R550)).^2;
R550=R550/max(max(R550));
R550=R550(384:639,384:639);
figure; imshow(R550);
title('Reconstructed image (550nm)')
Az3=Az3.*p;
S550=fftshift(fft2(fftshift(Az3)));
S550=(abs(S550)).^2;
S550=S550/max(max(S550));
S550=S550(384:639,384:639);

% Reconstruction (450nm)
w3=4500*10^-8;
H4=H2.*exp(2i*pi*sind(0.4)*(w-w3)/w/w3.*R*delta);
p=exp(-2i*pi*z.*((1/w3)^2-(1/M/delta)^2.*(C-M/2-1).^2-...
    (1/M/delta)^2.*(R-M/2-1).^2).^0.5);
Az4=fftshift(ifft2(fftshift(H4))).*p;
R450=fftshift(fft2(fftshift(Az4)));
R450=(abs(R450)).^2;
R450=R450/max(max(R450));
R450=R450(384:639,384:639);
figure; imshow(R450);
title('reconstructed image (450nm)')
Az4=Az4.*p;
S450=fftshift(fft2(fftshift(Az4)));
S450=(abs(S450)).^2;
S450=S450/max(max(S450));
S450=S450(384:639,384:639);

% color slit image
SLIT=zeros(256,256,3);
SLIT(:,:,1)=S650;
SLIT(:,:,2)=S550;
SLIT(:,:,3)=S450;
SLIT=uint8(SLIT.*500);
figure; image(SLIT)
title('Slit images')
axis off
axis equal
```

Problems

3.1 When the off-axis hologram is reconstructed with a light beam by the original reference wave [see Fig. 3.3(b)], the virtual image appears on-axis in the position of the original object and the real image is formed off-axis. Let us reconstruct the hologram with a light beam propagating at an angle $-\theta$ to the z-axis, i.e., we are using a conjugate beam to reconstruct the hologram. Show that the reconstructed real image is on-axis and the virtual image appears off-axis.

3.2 Verify that in lensless Fourier holography as discussed in Fig. 3.9, the reconstructed images are separated by $2fx_0 / z_0$, where f is the focal length of the lens used for reconstruction, x_0 is the separation distance between the spherical reference light and the object. In other words, you need to verify Eq. (3.16).

3.3 Consider an off-axis hologram. We assume that the object is at the optical axis and is z_0 behind the hologram. The offset angle of the plane wave reference light is θ_0, and the recording wavelength is λ_0. If the reconstruction light is also a plane wave with an offset angle θ_0 but with a wavelength of λ_r, show that the location of the reconstructed image is at $(z_r \sim (\lambda_0/\lambda_r)\, z_0;\, x_r \sim z_0\tan\theta_0(1 - \lambda_0/\lambda_r))$.

3.4 Assume that the slit for generating a rainbow hologram is z_0 from the hologram plane, and the offset angle of the plane wave reference light is θ_0. The recording wavelength is λ_0. With reference to Fig. 3.10(c), show that the relationship between the width of the slit, w, the diameter of the observer's pupil, d, and the bandwidth, $\Delta\lambda_0$, of the reconstructed image is given by $\Delta\lambda_0 \sim \lambda_0(w + d)/z_0\tan\theta_0$.

References

1. R. J. Collier, C. B. Burckhardt, and L. H. Lin, *Optical Holography* (Murray Hill, NJ, 1983).
2. P. Hariharan, *Optical Holography: Principles, Techniques, and Applications* (Cambridge University Press, Cambridge, 1996).
3. P. Hariharan, *Basics of Holography* (Cambridge University Press, Cambridge, 2002).
4. S. A. Benton, and V. M. Bove Jr., *Holographic Imaging* (Wiley, Hoboken, NJ, 2008).
5. D. Gabor, A new microscopic principle, *Nature* **161**, 777–778 (1948).
6. E. N. Leith, and J. Upatnieks, Reconstructed wavefronts and communication theory, *Journal of the Optical Society of America* **52**, 1123–1130 (1962).
7. S. A. Benton, Hologram reconstructions with extended incoherent sources, *Journal of the Optical Society of America* **59**, 1545–1546 (1969).
8. H. Chen, and F. T. S. Yu, One-step rainbow hologram, *Optics Letters* **2**, 85–87 (1978).
9. H. Chen, A. Tai, and F. T. S. Yu, Generation of color images with one-step rainbow holograms, *Applied Optics* **17**, 1490–1491 (1978).
10. J. C. Wyant, Image blur for rainbow holograms, *Optics Letters* **1**, 130–132 (1977).
11. E. N. Leith, and H. Chen, Deep-image rainbow holograms, *Optics Letters* **2**, 82–84 (1978).

4
Conventional digital holography

In Chapter 3 we introduced the fundamental theory of optical holography and reviewed several conventional holographic recording schemes. In digital holography, the recording schemes are the same, but the recording material is replaced by an electronic device, such as a charge-coupled device (CCD). Optical interference fringes acquired by the CCD are digitized into a two-dimensional digital signal and then processed using digital image processing to reconstruct the hologram. Figure 4.1(a) shows a typical case of digital holography and in this chapter we concentrate on this case. However, we may also come across other situations in digital holography. Complete numerical simulations of hologram construction and reconstruction can be performed purely by digital methods, as shown in Fig. 4.1(b).

We can also produce a digital hologram without any optical interference (i.e., a computer-generated hologram, see Chapter 7 for details). The hologram can then be sent to a display device for optical reconstruction, as shown in Fig. 4.1(c). In each of these situations, we need to simulate interference between the object and the hologram in the front end, and then simulate diffraction between the hologram and the diffraction plane or the observation plane in the back end of the overall process. Because creating the digital signal and digital calculations of diffraction are the core of digital holography, in this chapter we will first introduce the concept of discrete signals and their corresponding calculations for readers who are not familiar with digital signals and systems. Then we will discuss the properties and limitations of the recording device, i.e., a CCD. Finally, we will develop the discrete version of diffraction formulas and discuss any related problems.

4.1 Sampled signal and discrete Fourier transform

Details of digital signals and systems are beyond the scope of the book, so we will only discuss the most important properties that will be involved in digital

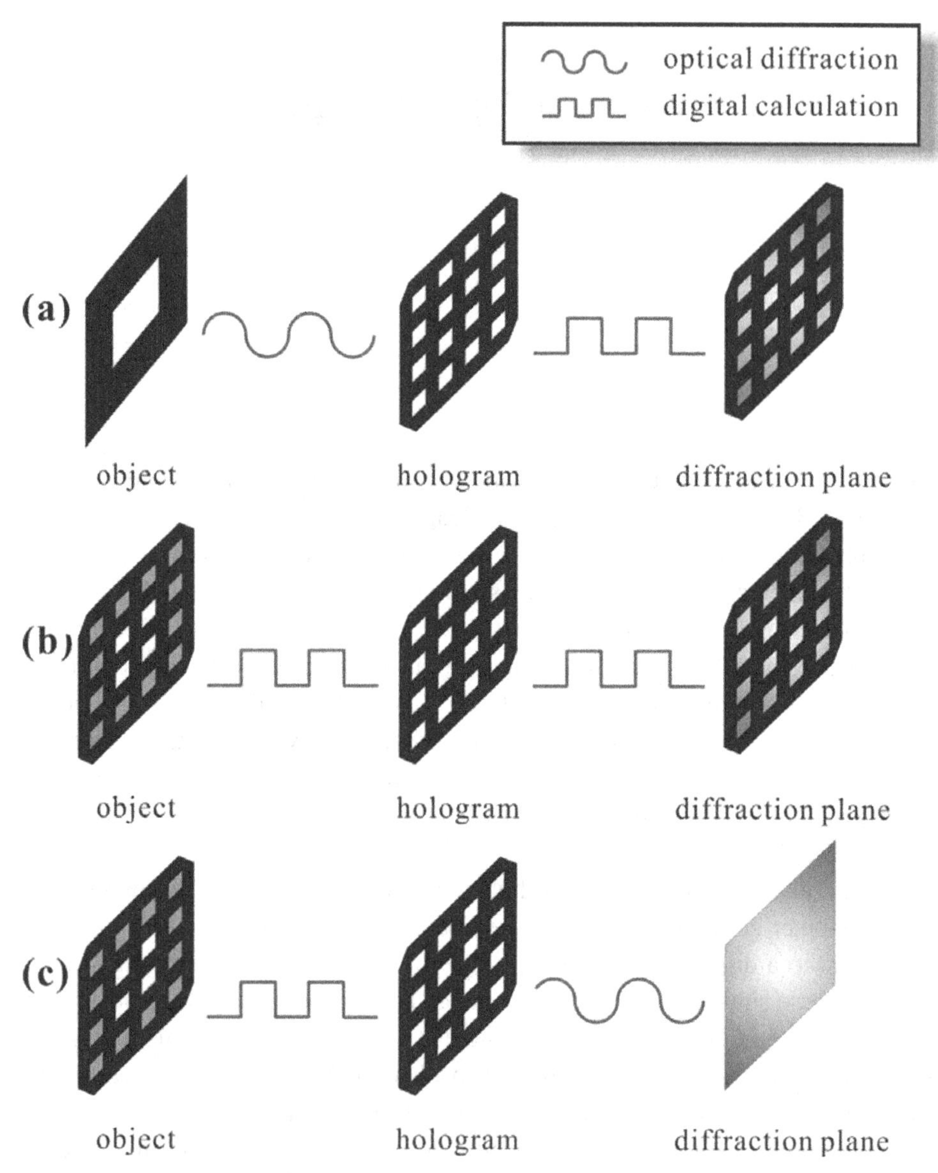

Figure 4.1 Three forms of digital holography.

holography. In the following discussion we treat only the one-dimensional case for simplicity without any loss of generality.

Assuming $f(x)$ is a continuous analog signal along the x-axis, we can produce a discrete signal corresponding to $f(x)$ by sampling it with a fixed separation, Δ_x,

$$f[n] = f(n\Delta_x) = \sum_{k=0}^{N-1} f(k\Delta_x) \times \delta([n-k]\Delta_x), \tag{4.1}$$

where n is an integer between 0 and $(N-1)$, and $\delta(n)$ is the *unit impulse sequence*, which is the counterpart of the delta function when dealing with discrete signals. The mathematical definition of $\delta([n-k]\Delta_x)$ is

$$\delta([n-k]\Delta_x) = \begin{cases} 1 & n = k \\ 0 & n \neq k. \end{cases} \tag{4.2}$$

In Eq. (4.1) we have also assumed that $f(x)$ is zero outside the range $0 \leq x < N\Delta_x$ so that we can use a finite number of samplings, say N, to represent the original analog signal. Because the original signal is sampled with *sampling period* Δ_x, the *sampling frequency* is $f_s = 1/\Delta_x$. In the following discussion we assume that N is even for the sake of simplicity, although it is not limited to being even. Also to simplify the discussion, we ignore the problem of finding the spectrum of a discrete signal. We must, however, know two important properties. First, the spectrum of the sampled signal $f[n]$ is periodic in the spectrum domain. This concept can be easily grasped in the analog version. By applying the Fourier transform to a continuous-defined sampled function, we get

$$\mathcal{F}\left\{f(x) \times \sum_{n=-\infty}^{\infty} \delta(x-n\Delta_x)\right\} = 2\pi f_s \sum_{n=-\infty}^{\infty} F(k_x - 2\pi f_s n), \tag{4.3}$$

where $F(k_x) = \mathcal{F}\{f(x)\}$. Thus the spectrum of the sampled function consists of the spectrum of the unsampled function which is replicated every $2\pi f_s$ radian/length. Assuming that the duplicates do not overlap with each other, the spectrum of the sampled function is periodic with period $2\pi f_s$, as shown in Fig. 4.2. So we can select any region of the spectrum with length $2\pi f_s$ to represent the complete spectrum information.

The interval over which the analog signal is sampled is referred to as the *record length*, L_x. With the number of samples N, the following relationships hold:

$$L_x = N\Delta_x = N/f_s \quad \text{and} \quad \Delta f = 1/L_x = f_s/N, \tag{4.4}$$

where Δf is *the frequency resolution*. In other words, for a given number of sample points, N, and sampling period, Δ_x, the sampling resolution in the frequency domain, Δf, is fixed. The resulting extent in the frequency domain is $L_f = 1/\Delta_x$.

To understand the above relationships, let us assume that the continuous signal $f(x)$ is sampled over the length $L_x = N\Delta_x$. Hence the spectrum of the continuous signal can be expressed as

$$\mathcal{F}\left\{f(x) \times \text{rect}\left(\frac{x}{L_x}\right)\right\} = F(k_x) * L_x \,\text{sinc}\left(\frac{k_x L_x}{2\pi}\right). \tag{4.5}$$

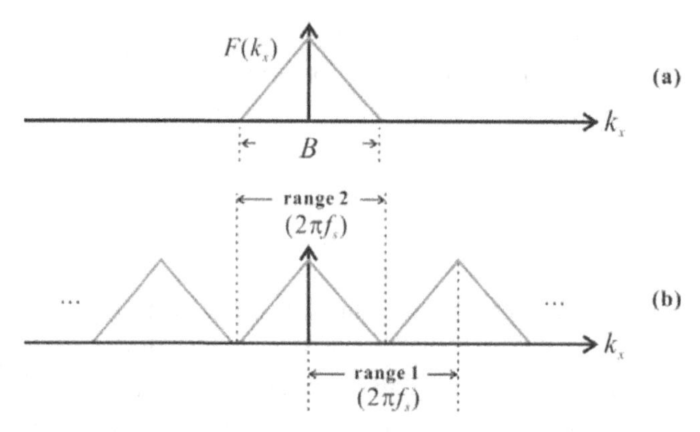

Figure 4.2 Spectrum of (a) the original continuous function, $f(x)$, and (b) the sampled function.

Since the resulting spectrum is the convolution of the original spectrum $F(k_x)$ and the sinc function, the sinc function can be regarded as an impulse response of a system with the original spectrum $F(k_x)$ treated as an input. Therefore the width of the impulse response defines the "frequency" resolution of the system. The "width," Δk, of $\mathrm{sinc}(k_x L_x/2\pi)$ can be defined by setting the argument of the sinc function to 1, i.e., the first zero of the sinc function, giving the width $\Delta k = 2\pi/L_x = 2\pi/N\Delta_x$ or $\Delta f = 1/N\Delta_x = 1/L_x$. So when the samples are Δf apart in the spectral domain, the value of Δf gives the frequency resolution of the resulting Fourier transform.

Now, for convenience, the frequency range of the spectrum is usually selected to be $[0, f_s)$ (this expression means 0 is included but f_s is not included) in the unit of cycle/length, or $[0, 2\pi f_s)$ in the unit of radian/length. Accordingly, the discrete spectrum $F[m]$ can be found from $f[n]$ by the *discrete Fourier transform (DFT)*,

$$F[m] = \sum_{n=0}^{N-1} f[n] \times \exp\left(-\frac{j2\pi nm}{N}\right), \tag{4.6a}$$

where m is an integer, being also between 0 and $(N - 1)$. Note that $F[m]$ is the N-point DFT of $f[n]$ and is itself a periodic sequence with period equal to N. Similar to the Fourier transform of a continuous signal, we can retrieve $f[n]$ from $F[m]$ by the *inverse discrete Fourier transform (IDFT)*,

$$f[n] = \frac{1}{N}\sum_{m=0}^{N-1} F[m] \times \exp\left(\frac{j2\pi nm}{N}\right). \tag{4.6b}$$

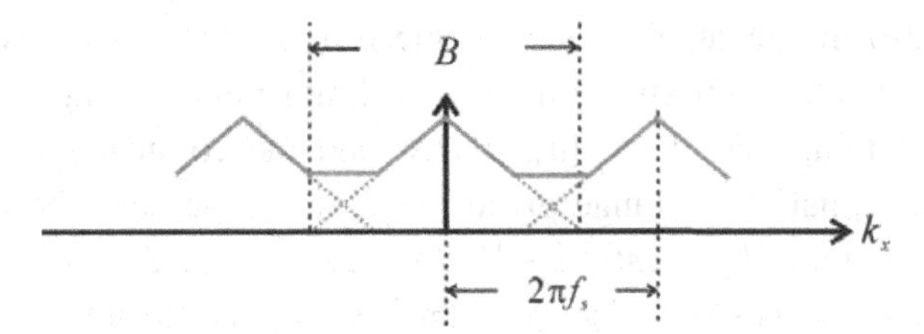

Figure 4.3 Spectrum of a sampled function when $B > 2\pi f_s$, illustrating spectral folding.

Note that $f[n]$ is periodic with period N samples or $N\Delta_x$ in the spatial domain. Similarly, the index m corresponds to the discretized radian frequencies k_m as $k_m\Delta_x = (2\pi/N)m$ if we recognize that $\exp(-j2\pi nm/N) = \exp(-jk_m n\Delta_x)$ from Eq. (4.6).

With reference to Fig. 4.2, we see that the spectrum range $[0, 2\pi f_s)$ is in "range 1." We observe that the right half of the central duplicate and the left half of its adjacent duplicate are included in range 1. As a result, twice the bandwidth of $F(k_x)$, B, must be smaller than $2\pi f_s$ if we want to avoid overlapping of the spectra. We define the *Nyquist frequency* f_{NQ} as

$$f_{NQ} = \frac{f_s}{2}, \tag{4.7}$$

and the maximum frequency of the signal must be smaller than f_{NQ}. If the duplicates overlap with each other, as shown in Fig. 4.3, there is no way to separate the two spectrum duplicates. In this case, the high frequency component is under-sampling and the retrieved spectrum is incorrect, resulting in errors called *aliasing* or *spectral folding*. In other words, in order to recover the original continuous signal faithfully from its sampled version, the continuous signal should be band-limited to $f_B = B/4\pi$ and the sampling frequency $f_s \geq 2f_B$. A signal is said to be band-limited to f_B if all of its frequency components are zero above f_B.

It should be noted that the DFT and IDFT [see Eq. (4.6)] are defined according to MATLAB, but not according to Eq. (1.22). In MATLAB, the DFT and IDFT can be evaluated by commands fft / ifft for one-dimensional functions and fft2 / ifft2 for two-dimensional functions, respectively. We use "Arial" font to represent MATLAB commands henceforth. These commands are named because of efficient algorithms known as *fast Fourier transform (FFT)* algorithms, which speed up the calculations of the DFT and IDFT, especially when the number of samples is large.

As shown in Fig. 4.2(b), for a well sampled signal the left half ($m = 0 \sim N/2 - 1$) of $F[m]$ is sufficient to reflect the spectrum properties of a real signal. In this case, the right half ($m = N/2 \sim N - 1$) of $F[m]$ is usually disregarded without any loss of information. However, in digital holography, a complex signal (i.e., complex amplitude of the optical field) is analyzed and thus not only the positive frequency spectrum but also the negative frequency spectrum of $F(k_x)$ is necessary. In other words, we prefer to obtain the spectrum in range 2 rather than in range 1 in Fig. 4.2(b). Because the sampling period of $f[n]$ is $1/\Delta_x$ and the command fft generates the spectrum of the signal in the range given by $[0, 2\pi f_s)$, we have the distribution of the spectrum in the range $k_m = 2\pi(N/2 - 1)/N\Delta_x \sim 2\pi(N - 1)/N\Delta_x$, which is the same as that in the range $k_m = 2\pi(-N/2 - 1)/N\Delta_x \sim 2\pi(-1)/N\Delta_x$. As a result, the full frequency spectrum between $\pm\pi f_s$ can be obtained directly by swapping the right and left halves of $F[m]$. After the swap, the spectrum is represented in $\pm\pi f_s$, or $(m = -N/2 \sim N/2 - 1)$, and the zero frequency $(m = 0)$ locates at the $(N/2 + 1)$ point of the swapped spectrum. In MATLAB, we can use the command fftshift to swap the data. Readers may also apply the command ifftshift, which is useful if the data number is odd.

Example 4.1: Under-sampling and aliasing

In this example we will see the effect of aliasing. We sample a sine function, $\sin(2\pi x/5)$, where x is in millimeters. The period of the function is 5 mm. According to the sampling theorem, the sampling period must be smaller than 2.5 mm in order to avoid aliasing. We use 0.1 mm, 0.2 mm, 0.4 mm, and 3 mm as the different sampling periods. The sampled data are then transformed to the spectral domain so that we can measure the frequency of the signal directly. In the continuous spatial domain, using a one-dimensional version of the two-dimensional Fourier transform [see Eq. 1.22)], we can find that

$$\mathcal{F}\{\sin(2\pi x/5)\} = j[\pi\delta(k_x + 2\pi/5) - \pi\delta(k_x - 2\pi/5)].$$

Therefore, we expect to have two "peaks" at $k_x = \pm 2\pi/5$ rad/mm or at $1/5 = 0.2\ \text{mm}^{-1}$ in the magnitude spectrum, i.e., $|\mathcal{F}\{\sin(2\pi x/5)\}|$, of the signal. The magnitude spectra are plotted in Fig. 4.4 (a) to (c) for the sampling periods of 0.1 mm, 0.2 mm, and 0.4 mm, respectively, and we observe that the two peaks in the spectrum are always at $0.2\ \text{mm}^{-1}$, which corresponds to the correct signal of period 5 mm. When the sampling period is 3 mm, the peaks of the spectrum are at $\pm 0.133\ \text{mm}^{-1}$ [Fig. 4.4(d)], which is incorrect due to aliasing. The MATLAB code is listed in Table 4.1.

Example 4.2: Sampling

We have an analog signal $f(x)$ that is bandlimited to 100 cycle/mm. The signal is to be sampled at the Nyquist frequency. The frequency resolution required is to be at 0.1 cycle/mm. Find the record length and the number of samples.

The Nyquist frequency is $f_{NQ} = 2 \times 100$ cycle/mm $= 200$ cycle/mm. To achieve a frequency resolution of 0.1 cycle/mm, the total record length is $L_x = 1/\Delta f = 1/0.1 = 10$ mm. The resulting number of samples is then $N = L_x f_{NQ} = 10 \times 200 = 2000$ samples.

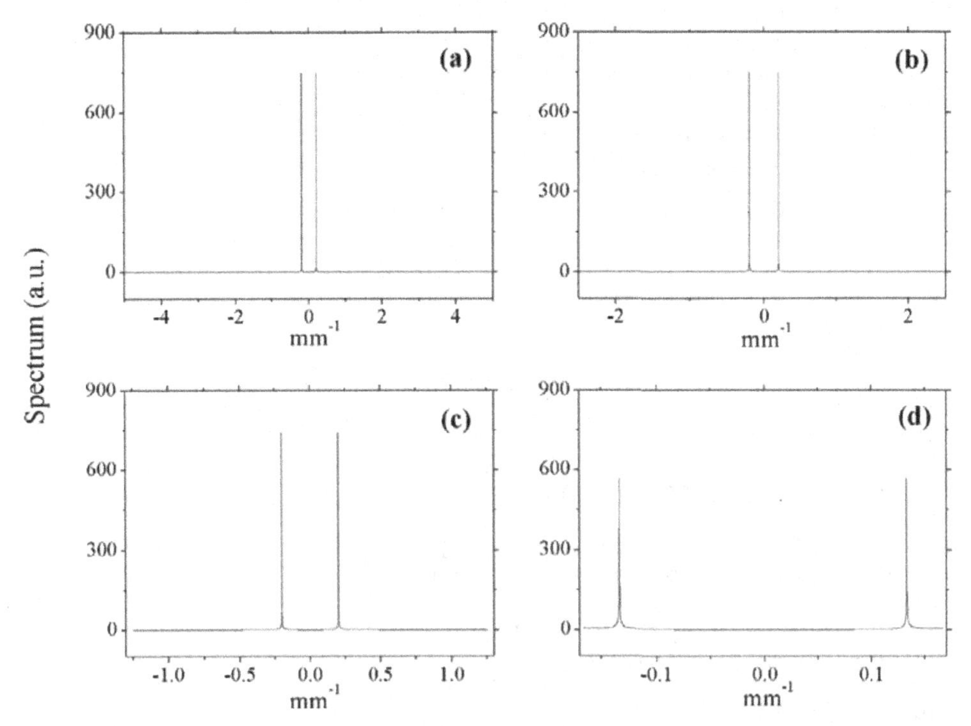

Figure 4.4 Magnitude spectra of a sine function sampled with periods at (a) 0.1 mm, (b) 0.2 mm, (c) 0.4 mm, and (d) 3 mm. The measured spatial frequencies are 0.2 mm^{-1} (a–c) and 0.133 mm^{-1} (d).

Example 4.3: Fourier transform of a rectangular function

We use a rectangular function and its Fourier transform as an example to show the implementation of the DFT using MATLAB. Table 4.2 is the MATLAB code and Fig. 4.5 shows the simulation results. At the first step, we center the aperture to the origin of the axis, which is at the $(N/2+1)$ point of $f[n]$. Then we apply command fftshift to $f[n]$ so that the origin of the axis is re-located to the first point, which is the requirement of the DFT (or IDFT) using MATLAB. Meanwhile, the $-x$ portion of the rectangular function is moved to $+x$. This can be done because the period of the phase in Eq. (4.6a) is N sampling points,

$$\exp\left[-\frac{j2\pi(n+N)m}{N}\right]=\exp\left[-\frac{j2\pi nm}{N}\right]\times\exp\left[-\frac{j2\pi Nm}{N}\right]=\exp\left[-\frac{j2\pi nm}{N}\right]. \quad (4.8)$$

Therefore in the DFT the phase contribution of the point at $-x$ is identical to that of the point at $(-x+N\Delta_x)$, which is what we do using fftshift. After the spectrum has been calculated using command fft, we perform fftshift again to re-locate the origin of the spectrum domain at the center of the axis, i.e., the $(N/2+1)$ point.

Table 4.1 *MATLAB code for demonstrating under-sampling and aliasing, see Example 4.1*

```
clear all; close all
x=0:0.1:150; % unit in mm sampling period 0.1 mm
y=sin(2*pi.*x/5); % period of 5mm
fmax=1/0.1/2;
fx=linspace(-fmax, fmax, 1501);
SY=abs(fftshift(fft(y)));

sx1=0:0.2:300; %unit in mm sampling period 0.2 mm
sy1=sin(2*pi.*sx1/5); % period of 5mm
fmax=1/0.2/2;
fx1=linspace(-fmax, fmax, 1501);
SY1=abs(fftshift(fft(sy1)));

sx2=0:0.4:600; %unit in mm sampling period 0.4 mm
sy2=sin(2*pi.*sx2/5); % period of 5mm
fmax=1/0.4/2;
fx2=linspace(-fmax, fmax, 1501);
SY2=abs(fftshift(fft(sy2)));

sx3=0:3:4500; %unit in mm sampling period 3 mm
sy3=sin(2*pi.*sx3/5); % period of 5mm
fmax=1/3/2;
fx3=linspace(-fmax, fmax, 1501);
SY3=abs(fftshift(fft(sy3)));

figure(1);
subplot(2,2,1)
plot(fx,SY)
axis([-inf, inf, 0, 800])
axis square
title('Sampling period 0.1 mm')
xlabel('1/mm')
ylabel('Spectrum')
subplot(2,2,2)
plot(fx1,SY1)
axis([-inf, inf, 0, 800])
axis square
title('Sampling period 0.2 mm')
xlabel('1/mm')
ylabel('Spectrum')
subplot(2,2,3)
plot(fx2,SY2)
axis([-inf, inf, 0, 800])
axis square
title('Sampling period 0.4 mm')
xlabel('1/mm')
ylabel('Spectrum')
subplot(2,2,4)
plot(fx3,SY3)
axis([-inf, inf, 0, 800])
axis square
title('Sampling period 3 mm')
xlabel('1/mm')
ylabel('Spectrum')
```

Table 4.2 *MATLAB code for calculating the Fourier spectrum of a rectangular function, see Example 4.3; the code is for plotting Fig. 4.5(a) and (b)*

```
clear all; close all;
N=64;                 %the total number of samples
L=10;                 %the half width of the rectangular function
dx=0.05;              %spatial resolution = 0.05 cm
x=-(N/2):(N/2)-1;%N samples centered at N/2+1
x=x* dx;
REC=zeros(1,N);
REC(1,(N/2+1-L):(N/2+L))=1;
%locate the aperture with width 2L to the center of the domain

SP=fft(fftshift(REC));
% first locate the center of the aperture to the fist point
% then calculate the spectrum of REC
SP=fftshift(SP);%locate the spectrum to the center of the domain
SP=SP/(max(SP)); % normalization
SP=real(SP);
df=1/N/dx; %frequency resolution
f=-(N/2):(N/2)-1;
f=f* df; %coordinate of spatial frequency centered at N/2+1
%produce the figures
figure(1)
subplot(1,2,1)
plot(x,REC)
axis([-inf, inf, -0.25, 1.1])
axis square
title('Rectangular function')
xlabel('cm')
subplot(1,2,2)
plot(f,SP)
axis([-inf, inf, -0.25, 1.1])
axis square
title('Spectrum')
xlabel('1/cm')
```

In Fig. 4.5(a), the spatial domain contains 64 points and the separation between adjacent sampled points is set to be $\Delta_x = 0.05$ cm. Hence according to Eq. (4.4), the total record length is $L_x = N\Delta_x = 64 \times 0.05 = 3.2$ cm and the frequency resolution is $\Delta f = 1/L_x = 0.3125$ cm^{-1} with the extent in the frequency domain being $L_f = 1/\Delta_x = 20$ cm^{-1}. The resulting spectrum is shown in Fig. 4.5(b). Note that the plot is jagged because the frequency resolution is poor. We can use *zero padding* to improve the frequency resolution, i.e., to pad zero-value points to the ends of the original spatial domain function. In Fig. 4.5(c), we keep the same $\Delta_x = 0.05$ cm but use 512 points (eight times as many points) to represent

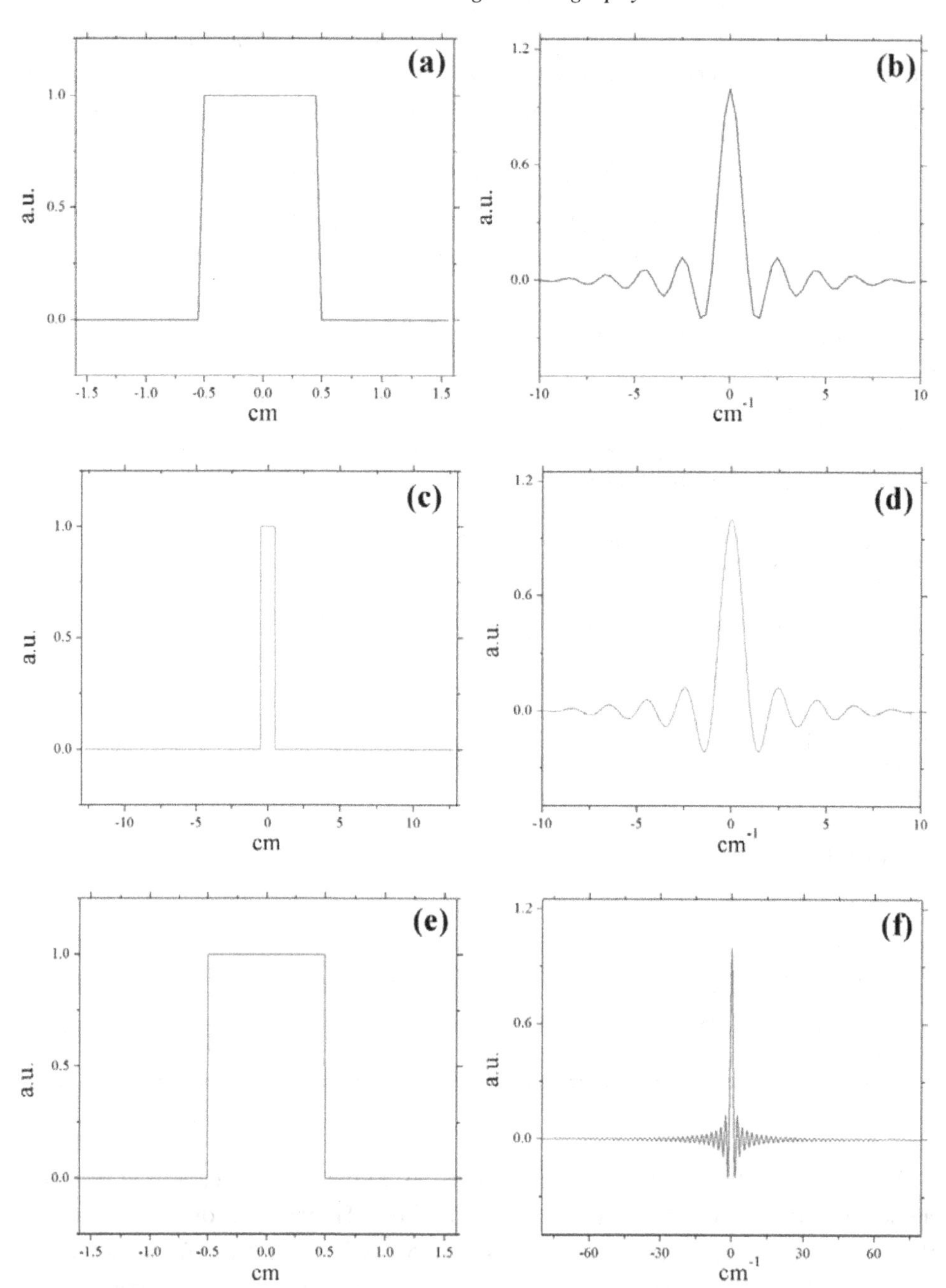

Figure 4.5 Rectangular function (left) and its Fourier spectrum (right). The spatial resolution and sample numbers of (a), (c), (e) are (Δ_x, N) = (0.05 cm, 64), (0.05 cm, 512), and (0.00625 cm, 512), respectively.

the whole spatial domain so that the domain size is eight times larger than that in Fig. 4.5(a), i.e., $L_x = N\Delta_x = 512 \times 0.05 = 25.6$ cm. The frequency resolution in the resulting spectrum becomes $\Delta f = 1/L_x = 0.039$ cm^{-1}, which is eight times better but with the same total length in the frequency domain as Δ_x is the same. The plot is shown in Fig. 4.5(d), which has a smoother appearance compared to the plot in Fig. 4.5(b).

Now in Fig. 4.5(e), we use 512 instead of 64 points to represent the same record length in Fig. 4.5(a), i.e., $L_x = N\Delta_x = 512 \times (\Delta_x/8) = 3.2$ cm. Although the sampling rate is eight times that of the original example, the sampling number is also eight times more than that of the original example. According to Eq. (4.4), the extent in the Fourier spectrum in Fig. 4.5(f) is eight times larger than that in Fig. 4.5(b), i.e., $L_f = 8/\Delta_x = 160$ cm^{-1}, but the frequency resolution is the same as that in Fig. 4.5(b) as $\Delta f = 1/L_x = 0.3125$ cm^{-1}. For a fixed spectrum range, say ± 10 cm^{-1}, the two curves in Fig. 4.5(b) and (f) are identical. In this case, more sampling points will not result in better precision in the frequency domain.

Finally, we want to point out a common misconception, i.e., that by padding zeros we improve accuracy. The truth is that once aliasing occurs, we can never obtain better accuracy by increasing the frequency resolution using zero padding. We can only increase the accuracy by not under-sampling. In our present example, we use a rectangular function, which is spatial-limited. Its Fourier transform is infinitely extended. Therefore, aliasing occurs no matter how small the frequency resolution achievable. Nevertheless, zero padding increases the number of samples and may help in getting a better idea of the spectrum from its samples. Zero padding is also useful to make up the required data points $N = 2^n$, i.e., being a power of 2 for efficient FFT operations.

4.2 Recording and limitations of the image sensor

The charge-coupled device (CCD) sensor and complementary metal-oxide-semiconductor (CMOS) sensor are two typical devices used to take digital photographs. Both these imagers are composed of numerous light sensing units, namely pixels, arranged as a two-dimensional array. The main difference between the CCD imager and the CMOS imager is that the signal from all the pixels shares one or several amplifiers in a CCD imager while each pixel has an independent amplifier in a CMOS imager. Although independent amplifiers allow the CMOS to process images with a faster frame rate, the properties of these amplifiers are not always uniform, resulting in slight distortion in the output image. Also, the dark current noise of the CMOS is larger than that of the CCD. Therefore the CCD is usually preferred in low-luminance applications or applications that require high image quality. The main advantage of the CMOS is that its structure is simple and thus

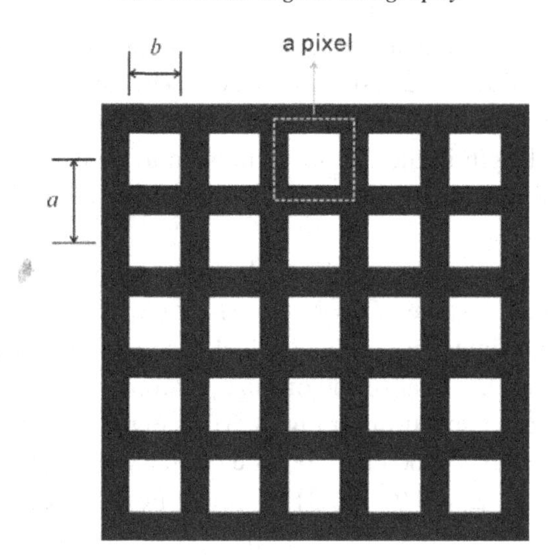

Figure 4.6 Schematic structure of a CCD/CMOS image sensor.

it is easily fabricated. CMOS imagers with more than ten thousand pixels in full-frame size (36 mm × 24 mm) have been used for commercial digital cameras. If the illumination is large enough, the CMOS imager can also provide high quality images. Therefore more and more digital holographic experiments have been demonstrated using CMOS imagers.

A CCD or a CMOS image sensor consists of a two-dimensional array of pixels, as shown in Fig. 4.6. In each pixel, there is shielding around the active area of the pixel (the white square in Fig. 4.6). The energies of photons impinging upon the active area of the pixel are transferred to electrons. Ideally the number of electrons in each pixel is proportional to the average intensity of the pixel. Because each pixel only delivers a single signal in a single illumination, the acquired image is a matrix with the same dimension as the imager. The pixel pitch, or the pixel size, is the distance between pixels ("a" in Fig. 4.6). The size of the imager, the pixel pitch and the fill factor, which is the ratio of active area $b \times b$ to the total area within a pixel (b^2/a^2), are three important parameters of an imager. For practical CCD sensors, the typical pixel pitch is 4~8 μm, and the fill factor is 80~90%. The common CCD sensor has 1024 × 768 pixels, so the chip size is about 6.0 × 4.5 mm^2. A perfect imager has to register a pattern whose value is proportional to the intensity on the CCD or CMOS chip. In digital holography, we also expect that the interference pattern can be recorded faithfully. However, the recording property is affected by the pixel pitch and other parameters of the imager. In the following paragraphs, we will discuss qualitatively the limitations of a practical imager.

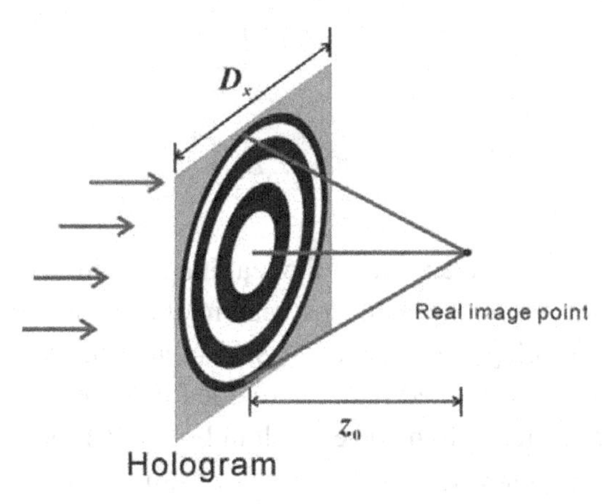

Figure 4.7 Reconstruction of an ideal FZP hologram with a square aperture.

4.2.1 Imager size

To see the effect of the finite extent of the imager, we consider a quasi-perfect imager in which the intensity profile can be recorded faithfully and continuously, but the size is limited. If we take a hologram of an infinitesimal point object using a normal incident plane wave reference light, then the resulting hologram will be a Fresnel zone plate (FZP), as we have shown in Section 2.2. As we reconstruct the hologram using a normal incident plane wave, we observe a real image point behind the hologram, as shown in Fig. 4.7. Obviously, the problem of holographic reconstruction is mathematically equal to the problem of focusing a collimated plane wave by a square-aperture ideal lens if we disregard the zeroth-order light and the twin image effects. The size of the real image point is $\sim\lambda_0 z_0/D_x$, where D_x is the extent of the imager along the x-direction, and z_0 is the distance between the point source and the recording medium in the recording stage. Thus the larger the imager size, the better the resolution.

4.2.2 Pixel pitch

Because the image acquired by an imager is discrete, i.e., the intensity profile is sampled as a digital signal, the practical resolution of the digitized image is usually below the diffraction limit, unless additional efforts are made. We employ an on-axis FZP to understand the situation. It is shown in Eq. (2.5) that the local fringe frequency of the FZP along the x-direction is $x/\lambda_0 z_0$, where again z_0 is the distance of the point source object from the holographic recording plane. For an imager with pixel pitch Δ_{xCCD} and according to the sampling theorem, in order to record the finest fringes at the edge, $D_x/2$, of the FZP, we must require the sampling frequency

$$f_s = 1/\Delta_{xCCD} \geq 2 \times (D_x/2)/\lambda_0 z_0 \tag{4.9}$$

or

$$z_0 \geq \frac{\Delta_{xCCD} D_x}{\lambda_0}. \tag{4.10}$$

Consequently, the reconstructed size of the real image point when the hologram is sampled by the imager is the same as that in the analog case, i.e., $\lambda_0 z_0/D_x$. Therefore, in digital holography the pixel size is also the minimum achievable resolved size of the reconstructed image, provided that the normal-incident plane wave is used as the reference light. The result in Eq. (4.10) shows that if the object distance, z_0, is smaller than $\Delta_{xCCD} D_x/\lambda_0$, the finer structures of the FZP cannot be resolved and aliasing occurs.

4.2.3 Modulation transfer function

We have just discussed the resolution limit due to the imager size and the sampling period. However, the image fidelity cannot be evaluated directly in the above analysis. Similar to a lens, the quality of an imager can be evaluated in terms of the modulation transfer function (MTF), which describes the relative frequency response of the output image of a system. The MTF of an imager mostly depends on the structure of the imager [1–4]. First, the periodic shielding structure (Fig. 4.6) can be regarded as a mask. Thus, presented in a one-dimensional case for simplicity, the effective illumination on the imager, $I'(x)$, can be written as

$$I'(x) = I(x) \times m(x), \tag{4.11}$$

where $I(x)$ is the illumination pattern on the imager and $m(x)$ is the intensity transmittance of the mask. Upon illumination of the active layer of the imager, photons are absorbed and electrons are excited simultaneously. Depending on the material used, the thickness of the active layer, and the wavelength of the illumination, excited electrons will laterally diffuse, resulting in an electron density $D(x)$ given by

$$D(x) = \beta I'(x) * G(x), \tag{4.12}$$

where $G(x)$ represents a Gaussian point spread function due to lateral diffusion, and β is a coupling constant, denoting the efficiency of the conversion of the electrons released by the light illumination. Finally, the signal from a single pixel of an imager is proportional to the integration of the illumination pattern on the pixel, which can be expressed as

$$\int_{x'-\frac{\gamma\Delta x_{CCD}}{2}}^{x'+\frac{\gamma\Delta x_{CCD}}{2}} D(x)dx, \tag{4.13}$$

where γ is a linear fill factor, which is defined by b/a, and the ratio is illustrated in Fig. 4.6; x' is the location of the considered pixel. Equation (4.13) can be re-written using the convolution operation as

$$\int_{-\infty}^{\infty} D(x) \times \text{rect}\left(\frac{x-x'}{\gamma\Delta_{xCCD}}\right) dx = D\left(x'\right) * \text{rect}\left(\frac{x'}{\gamma\Delta_{xCCD}}\right) = S\left(x'\right). \tag{4.14}$$

Thus for a one-dimensional image sensor containing N pixels, the nth pixel's output signal (n=1~N) can be expressed as

$$\begin{aligned} S[n] = S(n\Delta_{xCCD}) &= \left[D(n\Delta_{xCCD}) * \text{rect}\left(\frac{n\Delta_{xCCD}}{\gamma\Delta_{xCCD}}\right)\right] \\ &= \left\{\beta\left[I(n\Delta_{xCCD}) \times m(n\Delta_{xCCD})\right] * G(n\Delta_{xCCD}) * \text{rect}\left(\frac{n\Delta_{xCCD}}{\gamma\Delta_{xCCD}}\right)\right\}, \end{aligned}$$

or in analog form as

$$S(x) = \left\{\beta[I(x) \times m(x)] * G(x) * \text{rect}\left(\frac{x}{\gamma\Delta_{xCCD}}\right)\right\}. \tag{4.15}$$

For simplicity, we neglect the mask effect, i.e., letting $m(x) = 1$, so that we can define the modulation transfer function (MTF) as

$$\text{MTF} = |\mathcal{F}\{S(x)\}/\mathcal{F}\{\beta I(x)\}| = \left|\mathcal{F}\left\{G(x) * \text{rect}\left(\frac{x}{\gamma\Delta_{xCCD}}\right)\right\}\right|,$$

or

$$\text{MTF} = |\mathcal{F}\{G(x)\} \times \text{sinc}(\gamma\Delta_{xCCD}k_x/2\pi)|, \tag{4.16}$$

which shows that the rectangular function and $G(x)$ always act as lowpass filters for the illuminating pattern. For Si-based CCD imagers under short-wavelength illumination ($\lambda_0 < 0.8$ μm), the diffusion effect is minor [$G(x) \approx \delta(x)$] and the integrating effect of the pixel [Eq. (4.13)] dominates. A simulated MTF based on Eq. (4.16) ($\Delta_{xCCD} = 5$ μm; $\gamma = 0.9; \lambda_0 = 0.63$ μm) is plotted in Fig. 4.8. It shows that the MTF is not zero when the frequency is larger than the Nyquist frequency f_{NQ}, where f_{NQ} has been calculated according to Eq. (4.7) and

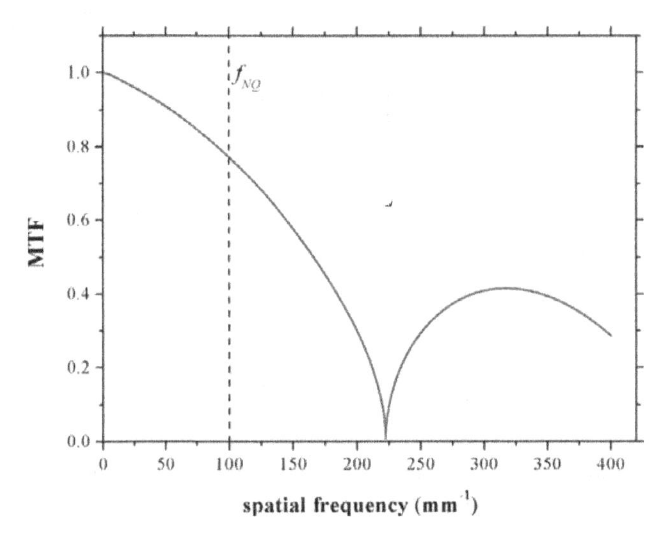

Figure 4.8 Simulated MTF curve for a CCD imager.

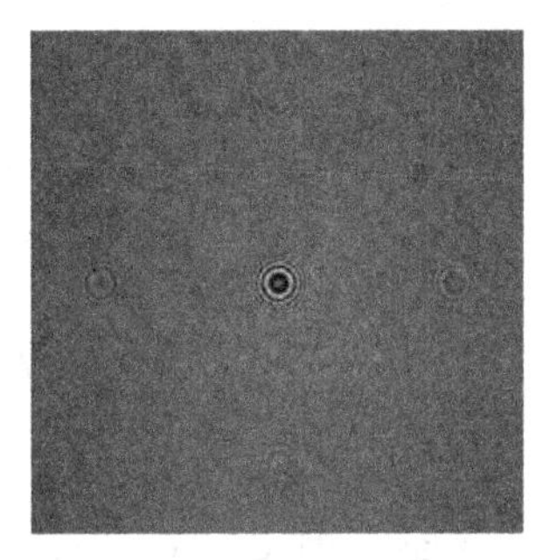

Figure 4.9 Photograph of a FZP taken by a CCD imager. Aliasing is clearly shown with the four faded circular patterns around the FZP.

is labeled in the figure. Aliasing will occur if the illumination pattern contains spatial frequencies beyond $f_{NQ} = 1/(2\Delta_{xCCD})$.

It is difficult to apply the advanced model, that is Eq. (4.15), to determine quantitatively the MTF of a specific CCD imager because we do not usually have enough information on $m(x)$ and $G(x)$. However, the MTF of a CCD imager can be measured by several experimental methods, which have been discussed in Refs. [5–7]. For general-purpose CCD imagers, high spatial frequency components are attenuated because of the roll-off of the MTF. Moreover, the response for frequencies larger than f_{NQ} is not zero, causing aliasing errors. Figure 4.9 shows a photograph of a FZP taken by a CCD imager (Kodak KAI-2020M). The central circular patterns are the true FZP structures while the darker circular patterns on the four sides are the false (aliasing) structures due to undersampling of fine fringes.

4.3 Digital calculations of scalar diffraction

4.3.1 Angular spectrum method (ASM)

In scalar diffraction theory [see Section 1.3], the diffracted field $\psi_p(x,y;z)$ can be obtained from the incident field, $\psi_{p0}(x, y)$, according to Eq. (1.27),

$$\psi_p(x, y; z) = \mathcal{F}^{-1}\left\{\mathcal{F}\left\{\psi_{p0}(x, y)\right\} \times \mathcal{H}(k_x, k_y; z)\right\}, \tag{4.17}$$

where $\mathcal{H}(k_x, k_y; z)$ is the *spatial frequency transfer function (SFTF)* [see Eq. (1.26)]. Equation (4.17) is the basic formula of the *angular spectrum method* (ASM), which is also called the *convolution method* or the *double Fourier transform method.*

Suppose that the sampling period along the x-axis is Δ_x with total M samples, and that the sampling period along the y-axis is Δ_y with total N samples, then the discrete version of Eq. (4.17) is expressed as

$$\psi_p[m, n] = \mathrm{DFT}_{2D}\left\{\mathrm{IDFT}_{2D}\left\{\psi_{p0}[m, n]\right\} \times \mathcal{H}[p, q]\right\}, \tag{4.18}$$

where

$$\mathcal{H}[p, q] = \exp\left[-jk_0 z\sqrt{1-\frac{(p\Delta_{kx})^2}{k_0^2}-\frac{(q\Delta_{ky})^2}{k_0^2}}\right] \tag{4.19}$$

and, according to Eq. (4.4), Δ_{kx} and Δ_{ky} are the frequency resolution (radian/unit of length) corresponding to sampling periods Δ_x and Δ_y along the x- and y-directions, respectively. Hence $\Delta_{kx} = 2\pi/M\Delta_x$ and $\Delta_{ky} = 2\pi/N\Delta_y$. In Eq. (4.18), (m, n) and (p, q) are the indices of the samples at the spatial domain and the Fourier domain, respectively. Thus their ranges are: $-M/2 \leq m \leq M/2 - 1$, $-N/2 \leq n \leq N/2 - 1$, $-M/2 \leq p \leq M/2 - 1$, and $-N/2 \leq q \leq N/2 - 1$. $\mathrm{DFT}_{2D}\{\bullet\}$ and $\mathrm{IDFT}_{2D}\{\bullet\}$ denote the two-dimensional discrete Fourier transform and the two-dimensional inverse discrete Fourier transform, respectively, and they are defined as

$$\mathrm{DFT}_{2D}\{f[m,n]\} = F[p,q] = \sum_{m=-\frac{M}{2}}^{\frac{M}{2}-1}\sum_{n=-\frac{N}{2}}^{\frac{N}{2}-1} f[m,n] \times \exp\left[-j2\pi\left(\frac{pm}{M}+\frac{qn}{N}\right)\right], \tag{4.20a}$$

$$\mathrm{IDFT}_{2D}\{F[p,q]\} = f[m,n] = \sum_{p=-\frac{M}{2}}^{\frac{M}{2}-1}\sum_{q=-\frac{N}{2}}^{\frac{N}{2}-1} F[p,q] \times \exp\left[j2\pi\left(\frac{pm}{M}+\frac{qn}{N}\right)\right]. \tag{4.20b}$$

Example 4.4: Diffraction of a rectangular aperture

In this example we simulate the diffraction of a rectangular aperture using the ASM. Table 4.3 gives the corresponding MATLAB code. In MATLAB, the

Table 4.3 *MATLAB code for calculating the diffraction field of a rectangular aperture using ASM, see Example 4.4*

```
clear all;close all;
lambda=0.633*10^-6; %wavelength, unit:m
delta=10*lambda; % sampling period, unit:m
z=0.2; % propagation distance; unit:m
M=512; % space size
c=1:M;
r=1:M;
[C, R]=meshgrid(c, r);
OB=zeros(M);
OB(246:267,246:267)=1;%create the rectangular aperture
SP_OB=fftshift(ifft2(fftshift(OB)));
deltaf=1/M/delta;%sampling period in the spectrum domain
c=1:M;
r=1:M;
[C, R]=meshgrid(c, r);
SFTF=exp(-2i*pi*z.*((1/lambda)^2-((R-M/2-1).*deltaf).^2-...
    ((C-M/2-1).*deltaf).^2).^0.5);
%spatial frequency transfer function of propagation of light
DF=fftshift(fft2(fftshift(SP_OB.*SFTF)));
Intensity=mat2gray((abs(DF)).^2);
figure; imshow(Intensity);
title('Intensity of the diffraction pattern')
```

command fftshift(fft2(fftshift(.))) performs $\mathrm{DFT}_{2D}\{\bullet\}$, and fftshift(ifft2(fftshift(.))) calculates $\mathrm{IDFT}_{2D}\{\bullet\}$. We perform fftshift before and after calculating fft2 (ifft2) in MATLAB because the origins in Eqs. (4.18), (4.19), and (4.20) are at the center of the coordinates. Also, the output from fft2 (ifft2) always gives the first point at the origin of the coordinates. Here we conduct two simulations, one for the propagation of 0.05 m and the other one for 0.2 m. The wavelength is 0.633 μm, and the size in the spatial domain is 512 × 512 pixels. The simulated diffraction patterns are shown in Fig. 4.10.

Note that because the size of the aperture (~0.14 mm) is much smaller than the propagation distance, the example can be regarded as diffraction of a rectangular aperture in the Fraunhofer region [Section 1.3.2]. So the intensity of the diffraction pattern should be sinc squared. However, we found that for a long propagation distance, grid-like artifacts appear in the diffraction pattern simulated using the ASM [Fig. 4.10(b)]. From the example, we know that the simulation results of Eqs. (4.18) and (4.19) are not always correct. In the following subsection we will explain why the simulations are sometimes incorrect. The validity of Eqs. (4.18) and (4.19) will be discussed as well.

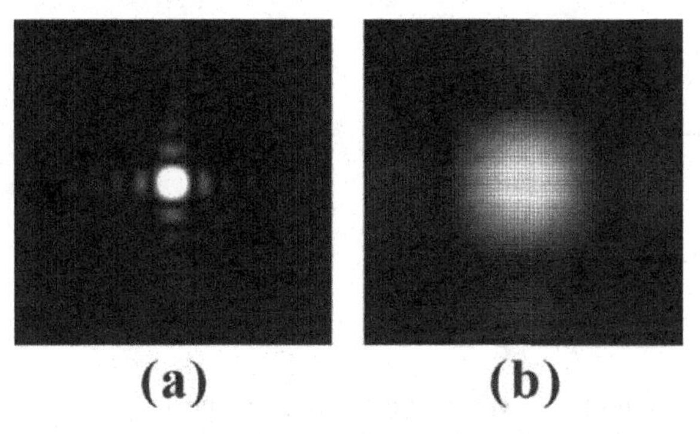

Figure 4.10 Intensity patterns of the diffracted field from a rectangular aperture simulated using the ASM when the propagation distance is (a) $z = 0.05$ m and (b) $= 0.20$ m.

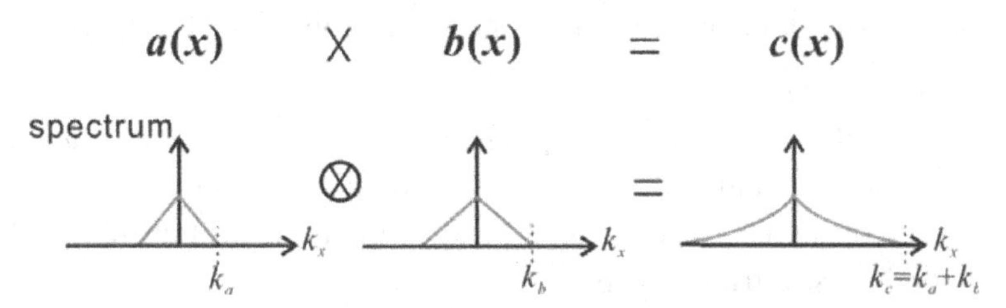

Figure 4.11 Product of two band-limited signals and spectrum of the resulting product.

4.3.2 Validity of the angular spectrum method

In Example 4.4, we showed that sometimes errors occur in the calculations using Eq. (4.18). In fact, the errors come from aliasing due to under-sampling. To explain the source of under-sampling, we first consider two arbitrary functions $a(x)$ and $b(x)$ band-limited to k_a and k_b, respectively. Then, what is the maximum spatial frequency of their product, $c(x) = a(x) \times b(x)$? With reference to Fig. 4.11, we see that the spectrum of $c(x)$ is equal to the convolution of the spectrum of $a(x)$ and that of $b(x)$. Consequently, $c(x)$ is band-limited to $k_c = k_a + k_b$. So even when the two functions, $a(x)$ and $b(x)$, are respectively sampled to $a[n]$ and $b[n]$ with their respective Nyquist frequencies, i.e., $2\pi f_{NQ,a} = 2k_a$ and $2\pi f_{NQ,b} = 2k_b$, we will undersample $c[n] = a[n] \times b[n]$ if employing similar sampling frequencies $f_{NQ,a}$ or $f_{NQ,b}$. Intuitively, we can increase the sampling rate of $a[n]$ and $b[n]$ to satisfy the well-sampling condition such that $2\pi f_{NQ,c} = 2k_c$ to avoid aliasing. This is the basic concept of what we will investigate in the following diffraction problem.

According to the above explanation, to convert Eq. (4.17) into Eq. (4.18), $\psi_{p0}(x, y)$ and $\mathcal{H}(k_x, k_y; z)$ must be well sampled. In addition, the product of $\mathcal{F}\left\{\psi_{p0}(x, y)\right\}$ and

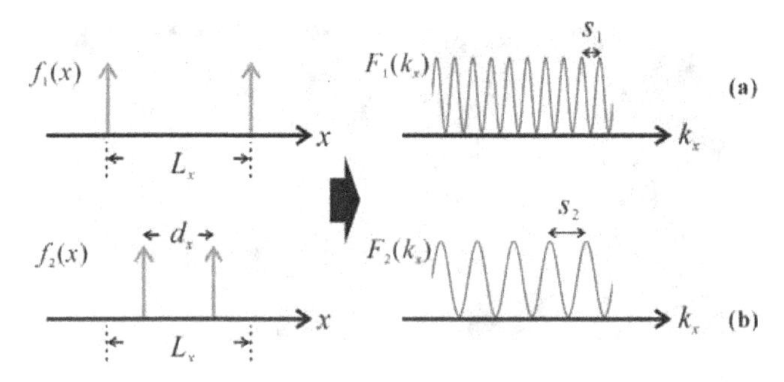

Figure 4.12 Example of two functions and their spectra.

$\mathcal{H}(k_x, k_y; z)$ must be well sampled in the spectral domain. Because the product is now in the spectral domain, we will analyze the properties of $\mathcal{F}\left\{\psi_{p0}(x, y)\right\}$ and $\mathcal{H}(k_x, k_y; z)$ separately.

The property of $\mathcal{F}\left\{\psi_{p0}(x, y)\right\}$ can be understood from the following examples where the signals contain only two pulses. In Fig. 4.12(a), we assume that the two pulses of signal $f_1(x)$ are at the edges of the measured range, L_x, and the corresponding spectrum will be

$$F_1(k_x) = \mathcal{F}\{\delta(x + L_x/2) + \delta(x - L_x/2)\} = 2 \cos\left(\frac{k_x L_x}{2}\right). \tag{4.21}$$

Therefore the "period" of $F_1(k_x)$ in the spectral domain is

$$s_1 = \frac{4\pi}{L_x} \text{ rad/length} \tag{4.22}$$

and the "frequency" of $F_1(k_x)$ can be expressed as

$$\frac{1}{s_1} = \frac{L_x}{4\pi} \text{ length/rad.} \tag{4.23}$$

Note that the period described in Eq. (4.22) is consistent with our earlier conclusion in Section 4.1 that the resolution element (a period contains two resolution elements) in the spectral domain is half of s_1, i.e., $2\pi/L_x$ with L_x being the record length in the spatial domain. In other words, the frequency resolution is $\Delta f = 1/M\Delta_x = 1/L_x$ in the spectral domain [Eq. (4.4)]. In Fig. 4.12(b), we assume another signal $f_2(x)$ also consisting of two pulses but with a smaller separation $d_x (d_x < L_x)$. Accordingly, the period of $F_2(k_x)$ is $s_2 = 4\pi/d_x$ and the frequency is $d_x/4\pi$. From these examples, we see that the larger the size of the signal in the spatial domain, the higher the "frequency" it contains in the spectral domain.

Now, let us consider a two-dimensional symmetric case ($\Delta_x = \Delta_y$ and $M = N$). We assume that for the incident field $\psi_{p0}(x, y)$, only the central $M' \times M'$ pixels contain non-zero values, where $0 < M' \leq M$. So according to Eq. (4.23), the frequency of the finest fringe of its spectrum is $M'\Delta_x/4\pi$. On the other hand, the local fringe frequency of $\mathcal{H}(k_x, k_y; z)$ (similar to the definition given in Eq. (2.5) but now in the spectral domain) can be found to be

$$\frac{d}{dk_r}\left[-k_0 z\sqrt{1-\frac{k_r^2}{k_0^2}}\right] = \frac{1}{2\pi}\frac{zk_r}{k_0\sqrt{1-k_r^2/k_0^2}}, \tag{4.24}$$

where $k_r = \sqrt{k_x^2 + k_y^2}$ and the local fringe frequency of $\mathcal{H}[p, q]$ along the horizontal or the vertical axis is at the maximum when

$$k_r = \Delta_{kx} \times M/2 = k_r^{max}, \tag{4.25}$$

where $\Delta_{kx} = 2\pi/M\Delta_x$. Consequently, according to Eq. (4.23), the maximum frequency of the product of $\mathcal{F}\left\{\psi_{p0}(x, y)\right\}$ and $\mathcal{H}(k_x, k_y; z)$ should be smaller than $M\Delta_x/4\pi$:

$$\frac{M'\Delta_x}{4\pi} + \frac{zk_r^{max}}{2\pi\sqrt{k_0^2-(k_r^{max})^2}} \leq \frac{M\Delta_x}{4\pi}. \tag{4.26}$$

From the above, we obtain

$$z \leq \frac{\sqrt{4\Delta_x^2-\lambda_0^2}}{2\lambda_0}(M-M')\Delta_x. \tag{4.27}$$

A correct simulation without the aliasing error can be obtained if the propagation distance satisfies Eq. (4.27). If the desired propagation distance is too long to satisfy Eq. (4.27), the sampling separation (Δ_x) can be increased, but the object spectrum must be shortened accordingly. Another way of satisfying Eq. (4.27) is to retain the object size (M') but use a larger pixel number (M). In other words, we can perform zero-padding prior to the simulation to avoid aliasing in the calculations of the diffracted field.

4.3.3 Fresnel diffraction method (FDM)

In this subsection we will introduce the Fresnel diffraction method (FDM), in which we apply the Fresnel diffraction formula to evaluate the diffracted field. The Fresnel diffraction formula, as introduced in Section 1.3.1, can be expressed using a single Fourier transform,

$$\psi_p(x, y; z) = \frac{jk_0 e^{-jk_0 z}}{2\pi z} e^{\frac{-jk_0}{2z}(x^2+y^2)} \times \mathcal{F}\left\{\psi_{p0}(x, y)\exp\left[\frac{-jk_0}{2z}(x^2+y^2)\right]\right\}\Bigg|_{\substack{k_x = \frac{k_0 x}{z} \\ k_y = \frac{k_0 y}{z}}} \cdot \tag{4.28}$$

To calculate Eq. (4.28) in the discrete space, we still assume that the sampling period along the x-axis is Δ_x with total M samples, and that the sampling period along the y-axis is Δ_y with total N samples. At the diffraction plane, the numbers of samples remain, but the sampling periods (Δ_x^d, Δ_y^d) change according to the following relationships:

$$\Delta_x^d = \frac{\lambda_0 z}{M\Delta_x}, \tag{4.29a}$$

$$\Delta_y^d = \frac{\lambda_0 z}{N\Delta_y}. \tag{4.29b}$$

Consequently, the discrete version of the Fresnel diffraction formula is given by

$$\psi_p[p, q] = \frac{jk_0 e^{-jk_0 z}}{2\pi z} Q_2[p, q] \times \mathrm{IDFT}_{2D}\left\{\psi_{p0}[m, n]Q_1[m, n]\right\}, \tag{4.30}$$

where

$$Q_1[m, n] = \exp\left[\frac{-jk_0}{2z}\left(m^2\Delta_x^2 + n^2\Delta_y^2\right)\right], \tag{4.31}$$

$$Q_2[p, q] = \exp\left\{-j\pi\lambda_0 z\left[\left(\frac{p}{M\Delta_x}\right)^2 + \left(\frac{q}{N\Delta_y}\right)^2\right]\right\} \tag{4.32}$$

with (m, n) and (p, q) the indices of the samples in the spatial domain (object plane) and in the Fourier domain (or diffraction plane), respectively. Thus their ranges are: $-M/2 \le m \le M/2 - 1$, $-N/2 \le n \le N/2 - 1$, $-M/2 \le p \le M/2 - 1$, and $-N/2 \le q \le N/2 - 1$. To obtain Eq. (4.30), we first replace (x, y) inside the transform by ($m\Delta_x$, $n\Delta_y$) to get $Q_1[m, n]$. We then replace (x, y) outside the transform by $(p\Delta_x^d, q\Delta_y^d)$ to get $Q_2[p, q]$ so that $\psi_p(x, y)$ becomes $\psi_p[p, q]$. In addition, it should also be noted that the IDFT is used in Eq. (4.30) instead of the DFT as the IDFT [see Eq. (4.6)] is defined according to MATLAB, but not according to Eq. (1.22). Note that we usually set $\Delta_x = \Delta_y$ and $M = N$ so that the sampling periods of the two axes on the diffraction plane are at the same scale. In the following subsection we consider this symmetrical case.

4.3.4 Validation of the Fresnel diffraction method

According to the discussion in Section 4.3.2, when we sample the product of $\psi_{p0}(x, y)$ and $\exp[(-jk_0/2z)(x^2+y^2)]$ to obtain $\psi_{p0}[m, n]Q_1[m, n]$ in Eq. (4.30), we need to satisfy the following relationship for a given sampling period Δ_x:

$$\text{highest frequency in } \psi_{p0}(x, y) + \text{highest frequency in } \exp\left[\frac{-jk_0}{2z}(x^2+y^2)\right] \leq \frac{1}{2\Delta_x}.$$

We can, for instance, assign half of the total bandwidth to each term on the left hand side of the above equation. There is, however, a z-dependence on $Q_1[m, n]$ and we need to determine the range of validity in z for a given bandwidth so as to avoid aliasing. $Q_1[m, n]$ is a quadratic-phase exponential function, and thus its spatial frequency can be evaluated in terms of the local fringe frequency. When evaluated at the edge of the field, i.e., $x = (M/2) \times \Delta_x$, the local fringe frequency should be smaller than $1/4\Delta_x$ (because it only shares half of the sampling bandwidth), that is,

$$\frac{(M\Delta_x/2)}{\lambda_0 z} \leq \frac{1}{4\Delta_x}, \tag{4.33a}$$

or

$$z \geq \frac{2M\Delta_x^2}{\lambda_0}. \tag{4.33b}$$

Therefore there is a minimum limit for the propagation distance to avoid the aliasing error.

In Eq. (4.30), $\psi_p[p, q]$ is proportional to the multiplication of the result of the inverse discrete Fourier transform and another quadratic-phase exponential function, $Q_2[p, q]$. The phase function $Q_2[p, q]$ can be ignored when the intensity distribution of the diffracted field is measured. For example, in the digital reconstruction of a hologram, we only take the modulus of the reconstructed field. If the phase of the diffracted field is of interest (e.g., holographic recording), $Q_2[p, q]$ cannot be ignored. Again, the maximum spatial frequency presented in this situation is the sum of the maximum spatial frequencies of $Q_2[p, q]$ and $\mathrm{IDFT}_{2D}\{\psi_{p0}[m, n]Q_1[m, n]\}$. Usually the edge of the output spatial domain is

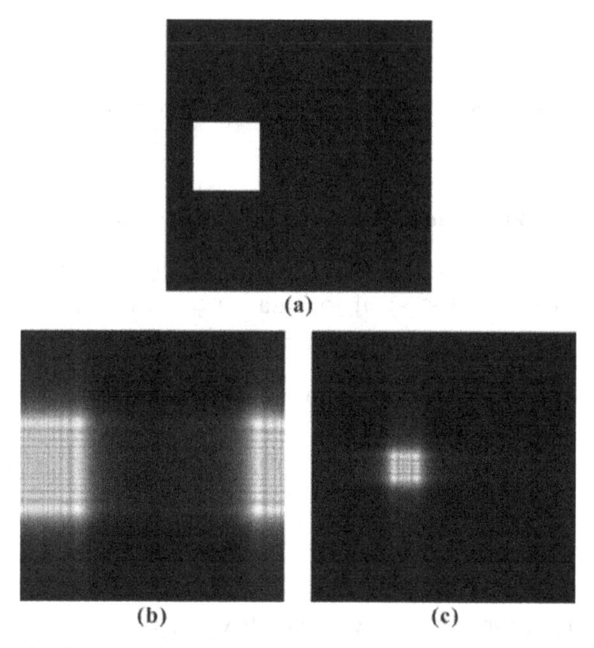

Figure 4.13 (a) Rectangular aperture, and modulus of the diffraction pattern at (b) $z = 0.02$ m and (c) $z = 0.06$ m.

under-sampled as the local frequency increases linearly away from the center of the hologram. A simple way to alleviate under-sampling is to interpolate the array of $\mathrm{IDFT}_{2D}\{\psi_{p0}[m, n]Q_1[m, n]\}$ before multiplying $Q_2[p, q]$. What is meant by interpolation, is to add more sampling points within the original adjacent sampling points. Thus the resulting sampling bandwidth can be improved. Another way is to perform zero-padding, that is, to pad zero pixels around the initial field before the calculation. However, the calculation load increases significantly using these two methods. Thus we recommend using the FDM to evaluate only the intensity distribution of the diffracted field for a propagation distance longer than $2M\Delta_x^2/\lambda_0$.

Example 4.5: Diffraction of a rectangular aperture (using FDM)

In this example we would like to simulate the diffraction of a rectangular aperture shown in Fig. 4.13(a). Table 4.4 gives the corresponding MATLAB code. We conduct two simulations for the propagation distances of 0.02 m and 0.06 m. Again, the wavelength is 0.633 μm, and the space size is 512×512 pixels. The results are shown in Fig. 4.13(b) and (c). In Fig. 4.13(b), the left side of the diffraction pattern moves to the right side of the window. This is due to under-sampling in the spatial domain. As the propagation distance is longer, i.e., satisfying Eq. (4.33b), this error disappears, as shown in Fig. 4.13(c).

Table 4.4 *MATLAB code for calculating the diffraction field of a rectangular aperture using the FDM, see Example 4.5*

```
close all; clear all;
lambda=0.633*10^-6; %wavelength, unit: m
delta=10*lambda; % sampling period, unit: m
z=0.06; % propagation distance; m
M0=512; % Object size
c=1:M0;
r=1:M0;
[C, R]=meshgrid(c, r);
THOR=((R-M0/2-1).^2+(C-M0/2-1).^2).^0.5;
A=THOR.*delta;
OB=zeros(M0);    %object pattern
OB(193:321,53:181)=1;
Q1=exp(-1i*pi/lambda/z.*(A.^2));
FTS=fftshift(ifft2(fftshift(OB.*Q1)));
%Fresnel diffraction

AFD=abs(FTS);
AFD=AFD/max(max(AFD));
figure; imshow(OB);
title('Rectangular aperture')
figure; imshow(AFD);
title('Modulus of the Fresnel diffraction pattern')
```

In Fig. 4.14, we plot the critical distances described by Eqs. (4.27) and (4.33b). In the plot, $M = 1024$ and $M' = 512$. It is apparent that, for any selected Δ_x, the ASM is valid for a shorter distance, while the FDM is valid for a longer distance. However, there is a forbidden gap, i.e., a region where aliasing occurs, between the two methods. For the ASM, the critical distance can be extended by increasing the spatial size M, i.e., zero-padding. For the FDM, the only way to shorten the critical distance is to reduce the sampling period because spatial size M cannot be smaller than M'.

4.3.5 Backward propagation

In the above discussion, we considered an aperture in the first plane (initial plane) and evaluated the diffracted field numerically in the second plane (diffraction plane). In the reconstruction of a digital hologram, we want to retrieve the object light in the initial plane from the second plane where the hologram is positioned. Thus we need to perform backward propagation from the hologram plane to the initial plane. We will see how to apply the ASM and the FDM to perform backward propagation.

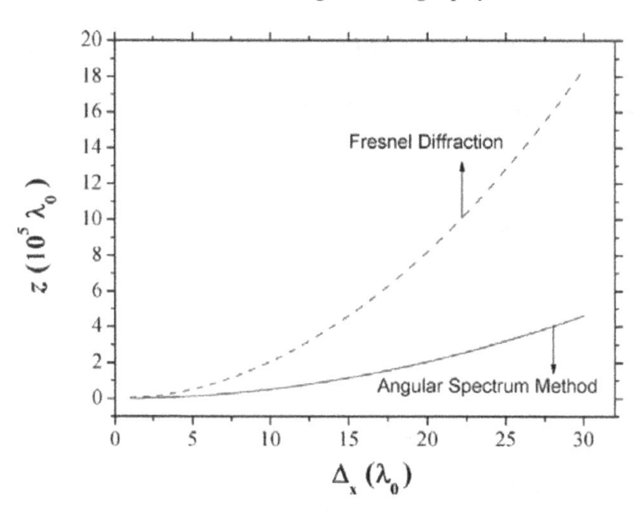

Figure 4.14 Critical distances for digital calculation using the angular spectrum method (ASM) and the Fresnel diffraction method (FDM).

For the AFM, we first Fourier transform and multiply $\mathcal{H}^{-1}$ on both sides of Eq. (4.17). We then write down a new form of Eq. (4.17):

$$\psi_{p0}(x,y)=\mathcal{F}^{-1}\left\{\mathcal{F}\left\{\psi_p(x,y;z)\right\}\times\mathcal{H}^{-1}(k_x,k_y;z)\right\}. \tag{4.34}$$

Because $\mathcal{H}^{-1}(k_x,k_y;z)=\mathcal{H}(k_x,k_y;-z)$, Eq. (4.34) can be expressed as

$$\psi_{p0}(x,y)=\mathcal{F}^{-1}\left\{\mathcal{F}\left\{\psi_p(x,y;z)\right\}\times\mathcal{H}(k_x,k_y;-z)\right\}. \tag{4.35}$$

So it is concluded that the ASM [Eq. (4.18)] can be applied directly to find the field at the initial field by setting a negative propagation distance. In backward propagation, the condition of alias-free simulation for the ASM [Eq. (4.27)] must also be satisfied, but now the z in Eq. (4.27) is replaced by $|z|$.

For the FDM, we can also re-arrange Eq. (4.30) to obtain the following form

$$\psi_{p0}[m,n]=\frac{-j2\pi z}{k_0}e^{jk_0z}Q_1^{-1}[m,n]\times\mathrm{DFT}_{2D}\left\{\psi_p[p,q]Q_2^{-1}[p,q]\right\}. \tag{4.36}$$

At a glance, Eq. (4.36) is different from Eq. (4.30). But it should be noticed that now we have the sampling periods in the diffraction plane (Δ_x^d,Δ_y^d), so we have to rewrite Eqs. (4.31) and (4.32) according to Eq. (4.29):

$$Q_2[p,q]=\exp\left[\frac{-jk_0}{2z}\left(p^2(\Delta_x^d)^2+q^2(\Delta_y^d)^2\right)\right], \tag{4.37}$$

$$Q_1[m,n] = \exp\left\{-j\pi\lambda_0 z\left[\left(\frac{m}{M\Delta_x^d}\right)^2 + \left(\frac{n}{N\Delta_y^d}\right)^2\right]\right\}. \tag{4.38}$$

Consequently, Eq. (4.36) can be re-expressed as

$$\psi_{p0}[m,n] = \frac{-j2\pi z}{k_0} e^{jk_0 z} Q_{2(-z)}[m,n] \times \text{DFT}_{2D}\left\{\psi_p[p,q] Q_{1(-z)}[p,q]\right\}, \tag{4.39}$$

where $Q_{2(-z)} = Q_{2(z)}^{-1}$ and $Q_{1(-z)} = Q_{1(z)}^{-1}$. In comparison with Eq. (4.30) and Eq. (4.39), we conclude that the FDM [Eq. (4.30)] can also be applied to find the intensity distribution in the initial plane. Neglecting the proportionality constant, the only difference is that in Eq. (4.30) $\text{IDFT}_{2D}\{\bullet\}$ is involved while in Eq. (4.39) $\text{DFT}_{2D}\{\bullet\}$ is involved. So in backward propagation using the FDM [Eq. (4.30)] with a negative propagation distance, the calculated intensity distribution will be inverted and flipped. One can invert the calculation result manually to show the correct diffraction pattern.

4.4 Optical recording of digital holograms

The recording of a digital hologram is simpler than recording on a holographic film. First, the sensitivity of a digital imager is several orders of magnitude higher than that of holographic film. Thus the problem of vibration during holographic recording in conventional holography is not serious in digital holography. Second, the response of a digital imager is linear so that problems of non-linearity associated with the film [9, 10] are negligible in digital holography. However, the main issue with digital holographic recording is that the maximum fringe frequency in the hologram should be smaller than the Nyquist frequency associated with the imager. Otherwise, the fringes are under-sampled and aliasing occurs.

4.4.1 Recording geometry

On-axis Fresnel holography

The bandwidth of the holographic fringes depends on the form of the reference light, the recording geometry, and the type of recording method. Here we take on-axis Fresnel holography with a plane wave reference wave as an example. Assume that the CCD contains $M \times M$ pixels, so its size is $M\Delta_{xCCD} \times M\Delta_{xCCD}$. The size of the object target is D and it is centered at the optical axis, as shown in Fig. 4.15. The distance between the object and the CCD sensor is z, and the plane wave reference wave is coupled to the object light via a beamsplitter (BS) or a half-silvered mirror. In the CCD plane, each object point together with the

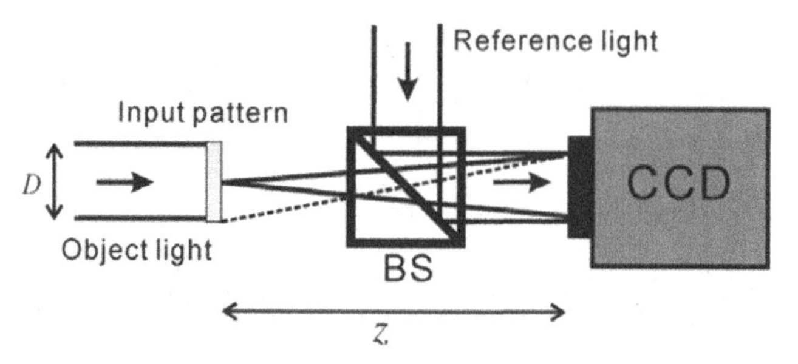

Figure 4.15 Setup of on-axis Fresnel holography.

reference wave forms a FZP. We can easily see that the ray of the object light emitting from its lower edge to the upper edge of the CCD sensor (dashed line in Fig. 4.15) forms a maximum angle with respect to the z-axis (rays forming diagonal connections between the object and the CCD sensor are ignored). Thus, the maximum transverse propagation distance of such a ray is $(D/2 + (M\Delta_{xCCD})/2)$, and the corresponding maximum local fringe frequency [see Eq. (2.5)] is $(D + M\Delta_{xCCD})/2\lambda_0 z$. To avoid under-sampling of the holographic fringes, the maximum local fringe frequency must be smaller than the Nyquist frequency [see Eq. (4.7)] associated with the CCD, that is

$$\frac{(D + M\Delta_{xCCD})}{2\lambda_0 z} \leq \frac{1}{2\Delta_{xCCD}}, \tag{4.40a}$$

or

$$z \geq \frac{\Delta_{xCCD}}{\lambda_0}(D + M\Delta_{xCCD}). \tag{4.40b}$$

Therefore, there is a minimum distance between the object and the CCD imager if aliasing is to be avoided.

Off-axis Fresnel holography

The above analysis can be extended to a generalized case in which the object pattern is not located at the center of the optical axis, and the plane wave reference wave is tilted, as shown in Fig. 4.16. The minimum distance is found to be

$$z \geq \frac{\Delta_{xCCD}(D + 2\delta + M\Delta_{xCCD})}{(\lambda_0 - 2\sin\theta\Delta_{xCCD})}, \tag{4.41}$$

where δ is the transverse shift of the object from its center, and θ is the offset angle of the reference wave. Note that in Eqs. (4.40) and (4.41), the zeroth-order beam

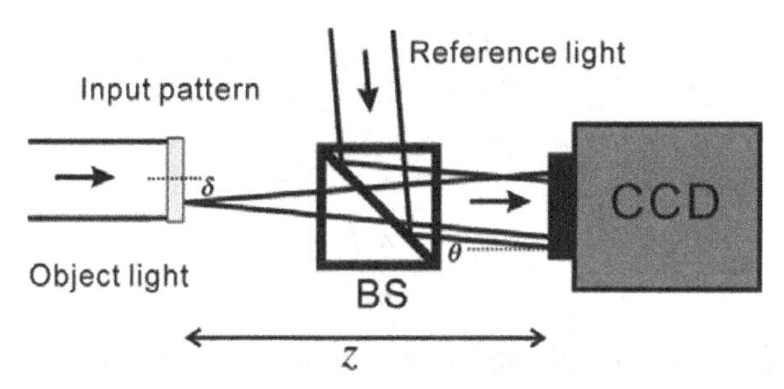

Figure 4.16 Setup of off-axis Fresnel holography.

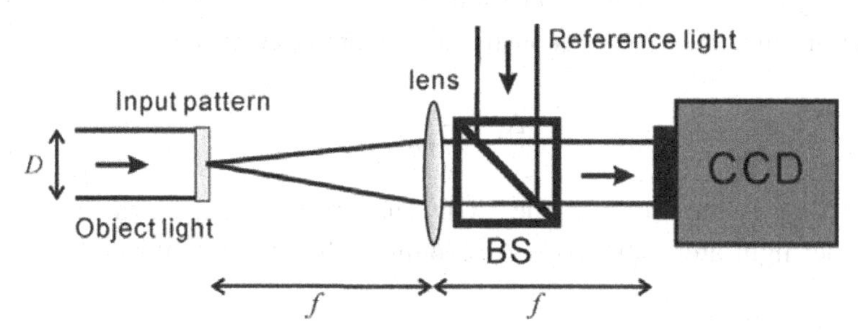

Figure 4.17 Setup of Fourier holography.

and the twin image are not considered. In other words, the condition in Eq. (4.41) ensures only that the hologram is well sampled and it does not guarantee that all these beams are separated in the reconstruction plane. We will discuss this aspect in Section 4.4.2.

Fourier holography

Figure 4.17 depicts the geometry of Fourier holography. Now the object and the CCD are at the front focal plane and the back focal plane of the Fourier transform lens, respectively. In this geometry, an off-axis object point produces a tilted plane wave on the CCD plane, resulting in a uniform sinusoidal fringe with the plane wave reference light [also see Fig. 3.7(a)]. However, the finest fringe is contributed by the point source emitting at the margin of the object at $x = D/2$. The finest fringe on the CCD plane therefore is given by [see Eq. (3.12)]

$$\left|1 + e^{jk_x D/2}\right|^2 = 2 + 2\cos\left[\pi \frac{xD}{\lambda_0 f}\right], \tag{4.42}$$

where $k_x = 2\pi x/\lambda_0 f$. Thus to resolve the finest fringe, the local fringe frequency must be smaller than the Nyquist frequency associated with the CCD, that is,

$$\frac{D}{2\lambda_0 f} \leq \frac{1}{2\Delta_{xCCD}}, \tag{4.43a}$$

or

$$D \leq \frac{\lambda_0 f}{\Delta_{xCCD}}, \tag{4.43b}$$

where f is the focal length of the Fourier transform lens. Note that the Fourier hologram can be regarded as a recording of the spectrum with resolution $\Delta_k = 2\pi\Delta_{xCCD}/\lambda_0 f$. So the size of the reconstruction space can be evaluated to be $2\pi/\Delta_k = \lambda_0 f/\Delta_{xCCD}$, which is consistent with Eq. 4.43(b). So the resolution, or the sampling distance Δ_x, of the reconstructed image is given by

$$\Delta_x = \frac{\lambda_0 f}{M\Delta_{xCCD}}, \tag{4.44}$$

where $M\Delta_{xCCD}$ is the width of the CCD imager. Again, the twin image and the zeroth-order light are not taken into account in the above analysis.

4.4.2 Removal of the twin image and the zeroth-order light

Fourier holography

Here we consider separating the zeroth-order light and the twin image from the reconstructed image in a Fourier hologram. The Fourier hologram is reconstructed by performing a single Fourier transform. The zeroth-order light is located at the center of the reconstruction plane, and the real image (the 1st order) appears at the location of the original object. The twin image (the −1st order) is flipped, inverted and a complex conjugate duplicate of the real image [see Fig. 3.7(b)]. Similar to the spectrum of the off-axis Fresnel hologram (Fig. 3.4), the sizes of the real image and the twin image are the same, while the size of the zeroth-order beam is twice that of the real image. Consequently, the size of the real image should be smaller than one fourth of the size of the reconstruction space to ensure that all the diffracted orders are separated on the reconstruction plane. In other words, in taking the Fourier hologram, the real image size as well as the object size should be smaller than $\lambda_0 f/4\Delta_{xCCD}$, according to Eq. (4.43b), and the object must be far from the center of the optical axis to avoid crosstalk between different diffracted orders.

Off-axis Fresnel holography

A single digital hologram acquired by a CCD imager is an amplitude hologram containing the real image, the virtual image, and the zeroth-order beam. Similar to optical holography, the three diffracted orders can be separated in the off-axis geometry (Fig. 4.16). In Section 3.2, we showed that the offset angle of the reference light should be large enough to avoid the interference between the 1st order and the zeroth-order beam, i.e.,

$$\frac{2\pi}{\lambda_0}\sin\theta \geq \frac{3}{2}B, \tag{4.45}$$

where B is the bandwidth of the 1st and -1st orders of light. On the other hand, the angle of the reference light cannot be too large. Otherwise, the 1st order will be under-sampled. Thus another condition is given by

$$\frac{2\pi}{\lambda_0}\sin\theta + \frac{B}{2} \leq \frac{2\pi}{2\Delta_{xCCD}}. \tag{4.46}$$

The bandwidth of the object light, B, depends on the light upon the CCD with the largest incident angle. Similar to the analysis of Eq. (4.40), the bandwidth can be determined to be

$$B = 2\pi\frac{(D + M\Delta_{xCCD})}{\lambda_0 z}. \tag{4.47}$$

Combining Eqs. (4.45), (4.46), and (4.47), the offset angle of the reference light is limited to the range

$$\sin^{-1}\left[\frac{3(D + M\Delta_{xCCD})}{2z}\right] \leq \theta \leq \sin^{-1}\left[\frac{\lambda_0}{2\Delta_{xCCD}} - \frac{(D + M\Delta_{xCCD})}{2z}\right]. \tag{4.48}$$

In Eq. (4.48), the critical condition occurs as the equals sign is selected, giving

$$\frac{3(D + M\Delta_{xCCD})}{2z} = \frac{\lambda_0}{2\Delta_{xCCD}} - \frac{(D + M\Delta_{xCCD})}{2z}. \tag{4.49}$$

After some manipulations, we can find the critical distance

$$z_c = \frac{4\Delta_{xCCD}(D + M\Delta_{xCCD})}{\lambda_0}, \tag{4.50}$$

which is also the minimum distance for successful off-axis holography. By substituting Eq. (4.50) into Eq. (4.48), we can also find the critical offset angle

$$\theta_c = \sin^{-1}\left(\frac{3\lambda_0}{8\Delta_{xCCD}}\right). \tag{4.51}$$

Equations (4.50) and (4.51) describe the optimized recording conditions for off-axis digital holography.

Example 4.6: Recording an off-axis digital hologram

We assume that a CCD contains 1024 × 1024 pixels with pixel pitch 8 μm. The size of the object to be recorded is 1 cm, and the wavelength is 0.633 μm. Calculate the optimized conditions for recording an off-axis digital hologram.

We can easily find the critical distance and the critical offset angle according to Eq. (4.50) and Eq. (4.51),

$$z_c = \frac{4(8\ \mu\text{m})(10^4\ \mu\text{m} + 1024 \times 8\ \mu\text{m})}{0.633\ \mu\text{m}} = 92\ \text{cm}$$

and

$$\theta_c = \sin^{-1}\left(\frac{3 \times 0.633\ \mu\text{m}}{8 \times 8\ \mu\text{m}}\right) = 1.7°.$$

So the offset angle is 1.7° and the minimum distance between the object and the CCD is 92 cm in order to avoid aliasing. It should be noted that according to Section 4.2, the resolution of the reconstructed image, using Eq. (4.50), now is

$$\frac{\lambda_0 z_c}{M\Delta_{xCCD}} = \frac{4(D + M\Delta_{xCCD})}{M} = \frac{4(10^4\ \mu\text{m} + 1024 \times 8\ \mu\text{m})}{1024} = 71\mu\text{m},$$

which is much larger than the pixel pitch of the CCD. If a larger offset angle as well as a larger propagation distance are selected, the resolution of the reconstructed image will be sacrificed further.

According to the above analysis, we can see that off-axis Fresnel holography is not a satisfactory method for acquiring digital holograms because most of the bandwidth of the imager is wasted. In Fourier holography, the bandwidth of the reconstructed real image only depends on the size of the imager. However, the object size is limited to one fourth of the field of view. Phase-shifting holography (PSH) is an efficient technique to alleviate the problem of the twin image and the zeroth-order light without sacrificing the bandwidth or the field of view. In PSH, several digital holograms with different phase shifts between them are used to retrieve the complex field of the object light. The details of PSH will be discussed in Section 5.1.

4.5 Simulations of holographic recording and reconstruction

In this section, we simulate the recording and the reconstruction of a Fresnel digital hologram. The complete simulation can be separated into three steps.

Step 1: Find the field in the CCD plane (hologram plane) from the field in the object plane.
Step 2: Introduce a reference light to the CCD plane and calculate the resulting interference fringes. This step together with step 1 is to simulate holographic recording.
Step 3: Find the field in the object plane from the field producing from the hologram (backward propagation). This step is to simulate holographic reconstruction.

In the following two examples we will conduct simulations of on-axis Fresnel holography and off-axis Fresnel holography.

Example 4.7: Simulation of on-axis holographic recording and reconstruction

In this example we would like to simulate on-axis holographic recording and reconstruction. The simulation parameters are set as follows. The size of the input

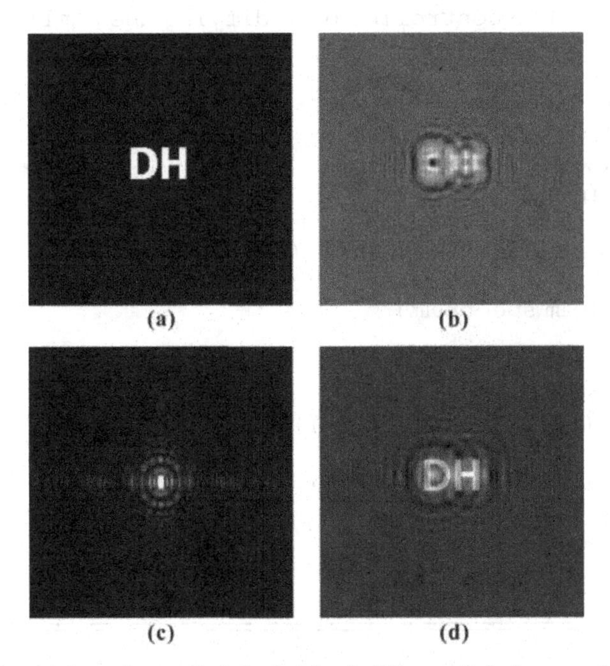

Figure 4.18 (a) Input image (original object), (b) on-axis hologram, (c) spectrum of (b), and (d) reconstructed image.

Table 4.5 *MATLAB code for simulation of on-axis holographic recording and reconstruction, see Example 4.7*

```
clear all; close all;
I=imread('DH256.bmp' , 'bmp');% 256×256 pixels, 8bit image
I=double(I);
% parameter setup
M=256;
deltax=0.001; % pixel pitch 0.001 cm (10 um)
w=633*10^-8; % wavelength 633 nm
z=25; % 25 cm, propagation distance

%Step 1: simulation of propagation using the ASM
r=1:M;
c=1:M;
[C, R]=meshgrid(c, r);
A0=fftshift(ifft2(fftshift(I)));
deltaf=1/M/deltax;
p=exp(-2i*pi*z.*((1/w)^2-((R-M/2-1).*deltaf).^2-...
    ((C-M/2-1).*deltaf).^2).^0.5);
Az=A0.*p;
EO=fftshift(fft2(fftshift(Az)));

%Step 2: interference at the hologram plane
AV=4*(min(min(abs(EO)))+max(max(abs(EO))));
%amplitude of reference light
%visibility can be controlled by modifying the amplitude
IH=(EO+AV).*conj(EO+AV);
figure; imshow(I);
title('Original object')
axis off
figure; imshow(mat2gray(IH));
title('Hologram')
axis off
SP=abs(fftshift(ifft2(fftshift(IH))));
figure; imshow(500.*mat2gray(SP));
title('Hologram spectrum')
axis off

%Step 3: reconstruction
A1=fftshift(ifft2(fftshift(IH)));
Az1=A1.*conj(p);
EI=fftshift(fft2(fftshift(Az1)));
EI=mat2gray(EI.*conj(EI));
figure; imshow(EI);
title('Reconstructed image')
axis off
```

image is 256 × 256 pixels; the distance between the object and the CCD is 25 cm; the pixel pitch is 10 μm, and the wavelength is 0.633 μm. For simplicity, we assume that the object is at the central area of the input image so that we can ignore the procedure of zero-padding in the simulation of propagation. The simulation result is shown in Fig. 4.18, while the MATLAB code is listed in Table 4.5. Note that in Fig. 4.18(d) the reconstructed image is disturbed by the zeroth-order light and the twin image.

We know that the zeroth-order light and the twin image can be separated from the reconstructed image using off-axis holography. The code listed in Table 4.5 can be applied directly for the simulation of off-axis holography. However, the result is noisy due to aliasing issues. So the simulation method must be modified for off-axis holography.

First in step 1, we need to know the phase distribution on the hologram plane. We usually use the ASM (Sections 4.3.1 and 4.3.2) to simulate light propagation. However, the condition in Eq. (4.27) must be satisfied in the simulation to avoid aliasing. The size of the CCD (hologram) may be larger than, equal to, or smaller than the workspace. Thus we can crop the resulting field or pad zeros around the resulting field to fit the size of the hologram. For example, if the hologram size is 1024 × 768, we can crop the central 512 × 512 pixels of the hologram for reconstruction. Then the size of workspace in this case is 512 × 512.

In step 2, we add a reference light to the object light and take the intensity of the superposed fields. The bandwidth of the workspace must be twice the bandwidth of the object light if the hologram is an on-axis hologram. If the hologram is an off-axis hologram, the bandwidth of the workspace must be at least four times the bandwidth of the object. We can interpolate the field data obtained in step 2 to ensure sufficient bandwidth.

In step 3, we usually apply FDM (Sections 4.3.3 and 4.3.4) to simulate reconstruction, provided the phase of the reconstructed image is not of interest. So the condition in Eq. (4.33b) must be satisfied in the simulation.

Because proper zero-padding must be applied in the simulation of propagation, the workspace grows step by step. So we crop the resulting field or hologram at the end of every step to reduce the computation loading. Example 4.8 is a demonstration of the simulation in off-axis holography.

Example 4.8: Simulation of off-axis holographic recording and reconstruction

The simulation parameters are set as follows. The size of the input image is 256 × 256 pixels; the distance between the object and the CCD is 20 cm; the pixel pitch is 10 μm, and the wavelength is 0.633 μm. The MATLAB code is listed in Table 4.6, while the results are shown in Fig. 4.19.

Table 4.6 *MATLAB code for simulation of off-axis holographic recording and reconstruction, see Example 4.8*

```
clear all; close all;
%%Reading input bitmap file
I0=imread('DH256.bmp', 'bmp');%256x256 pixels, 8bit image
I0=double(I0);
% parameter setup
M=256;
deltax=0.001; % pixel pitch 0.001 cm (10 um)
w=633*10^-8; % wavelength 633 nm
z=20; % z=M*deltax^2/w; % propagation distance

% Step 1: simulation of propagation
r=1:5*M;
c=1:5*M;
[C, R]=meshgrid(c, r);
I=zeros(5*M);
I(513:768,513:768)=I0;
A0=fftshift(ifft2(fftshift(I)));
deltaf=1/5/M/deltax;
p=exp(-2i*pi*z.*((1/w)^2-((R-641).*deltaf).^2-...
    ((C-641).*deltaf).^2).^0.5);
Az=A0.*p;
EO=fftshift(fft2(fftshift(Az)));
EO=EO(513:768,513:768);%reduce diffraction-plane size

% Step 2: interference at the hologram plane
% zero-padding in the spectrum domain
Az=fftshift(ifft2(fftshift(EO)));
Az2=zeros(4*M);
Az2(385:640,385:640)=Az;
EOf=fftshift(fft2(fftshift(Az2)));

AV=(min(min(abs(EOf)))+max(max(abs(EOf))))/2;
angle=0.3; % reference beam angle; degree
r2=1:4*M;
c2=1:4*M;
[C2, R2]=meshgrid(c2, r2);
Ref=AV*exp(1i*2*pi*sind(angle)*deltax/4.*...
    (R2-2*M-1)/w+1i*2*pi*sind(angle)*deltax/4.*...
    (C2-2*M-1)/w);
IH=(EOf+Ref).*conj(EOf+Ref);
IH=IH(257:768,257:768);%reduce the hologram size
figure; imshow(mat2gray(IH));
title('Hologram')
axis off
SP=fftshift(ifft2(fftshift(IH)));
```

Table 4.6 (*cont.*)

```
figure; imshow(50.*mat2gray(abs(SP)));
title('Hologram spectrum')
axis off

% Step 3: reconstruction (Fresnel diffraction)
r3=1:2*M;
c3=1:2*M;
[C3, R3]=meshgrid(c3, r3);
THOR=((R3-M-1).^2+(C3-M-1).^2).^0.5;
A=THOR.*deltax/4;
QP=exp(1i*pi/w/z.*(A.^2));
FTS=fftshift(fft2(fftshift(IH.*QP)));
I2=FTS.*conj(FTS);
figure; imshow(5.*mat2gray(I2));
title('Reconstructed image')
axis off
```

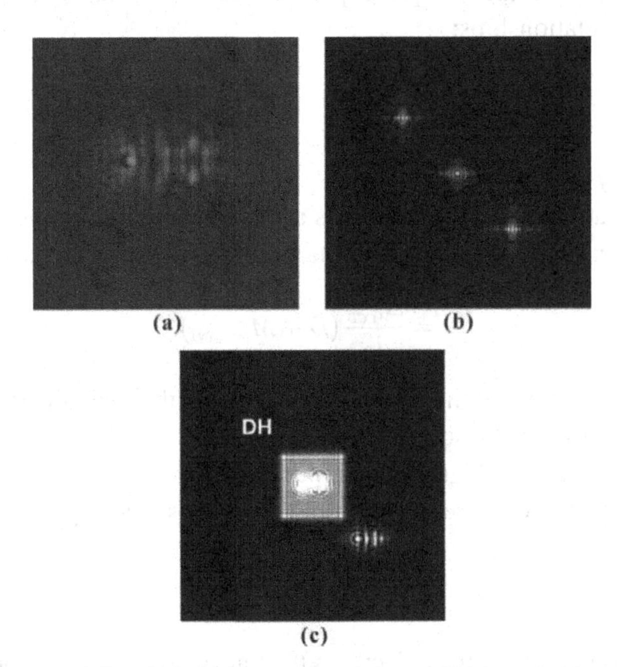

Figure 4.19 (a) Off-axis hologram, (b) spectrum of (a), and (c) reconstructed image.

Problems

4.1 Following Example 4.3, calculate and plot the spectrum of a shifted rectangular function, i.e., the rectangular function is not centered at the axis origin.

4.2 Perform and plot the FFT of a rectangular function, rect(x), using MATLAB. Also plot the spectrum of the rectangular function using the result from Table 1.1. Compare the two plots and if you are performing the calculations correctly, you will see that there is a little difference between the two curves.

4.3 Show that the spectrum of the sampled signal $f(x)\times\sum_{n=-\infty}^{\infty}\delta(x-n\Delta_x)$ is periodic in the spectrum domain as follows:

$$\mathcal{F}\left\{f(x)\times\sum_{n=-\infty}^{\infty}\delta(x-n\Delta_x)\right\}=2\pi f_s\sum_{n=-\infty}^{\infty}F(k_x-2\pi f_s n),$$

where $f_s=1/\Delta_x$.

4.4 Using the Fresnel diffraction method (FDM) to calculate the diffracted field, and if the sampling period on the object plane along the x-axis is Δ_x with total M samples, and the sampling period along the y-axis is Δ_y with total N samples, show that at the diffraction plane with the number of samples being the same, the sampling periods (Δ_x^d,Δ_y^d) change according to the following relationships:

$$\Delta_x^d=\frac{\lambda_0 z}{M\Delta_x},\qquad \Delta_y^d=\frac{\lambda_0 z}{N\Delta_y},$$

as given by Eq. (4.29).

4.5 Assume that an on-axis hologram is acquired according to the well-sampling condition in Eq. (4.40), i.e., the distance between the object and the CCD is

$$z=\frac{\Delta_{xCCD}}{\lambda_0}\left(D+M\Delta_{xCCD}\right).$$

Ignore the problem of the zeroth-order light and the twin image. Calculate the resolution of the reconstructed image.

4.6 In off-axis Fresnel holography and for aliasing-free calculations, show that the minimum distance of the object from the CCD is [see Eq. (4.41)]

$$z_{min}\geq\frac{\Delta_{xCCD}\left(D+2\delta+M\Delta_{xCCD}\right)}{\left(\lambda_0-2\sin\theta\Delta_{xCCD}\right)},$$

where D is the size of the object, δ is the transverse shift of the object from the center of the object plane, and θ is the offset angle of the reference wave as shown in Fig. 4.16.

References

1. D. H. Seib, Carrier diffusion degradation of modulation transfer function in charge coupled imagers, *IEEE Transactions on Electron Devices* **21**, 210–217 (1974).
2. S. G. Chamberlain, and D. H. Harper, MTF simulation including transmittance effects and experimental results of charge-coupled imagers, *IEEE Transactions on Electron Devices* **25**, 145–154 (1978).
3. J. C. Feltz, and M. A. Karim, Modulation transfer function of charge-coupled devices, *Applied Optics* **29**, 717–722 (1990).
4. E. G. Stevens, A unified model of carrier diffusion and sampling aperture effects on MTF in solid-state image sensors, *IEEE Transactions on Electron Devices* **39**, 2621–2623 (1992).
5. M. Marchywka, and D. G. Socker, Modulation transfer function measurement technique for small-pixel detectors, *Applied Optics* **31**, 7198–7213 (1992).
6. A. M. Pozo, and M. Rubiño, Comparative analysis of techniques for measuring the modulation transfer functions of charge-coupled devices based on the generation of laser speckle, *Applied Optics* **44**, 1543–1547 (2005).
7. A. M. Pozo, A. Ferrero, M. Ubiño, J. Campos, and A. Pons, Improvements for determining the modulation transfer function of charge-coupled devices by the speckle method, *Optics Express* **14**, 5928–5936 (2006).
8. J.-P. Liu, Controlling the aliasing by zero-padding in the digital calculation of the scalar diffraction, *Journal of the Optical Society of America A* **29**, 1956–1964 (2012).
9. A. A. Friesem, and J. S. Zelenka, Effects of film nonlinearities in holography, *Applied Optics* **6**, 1755–1759 (1967).
10. O. Bryngdail, and A. Lohmann, Nonlinear effects in holography, *Journal of the Optical Society of America* **58**, 1325–1330 (1968).

5
Digital holography: special techniques

In Chapter 4, we introduced conventional digital holography, describing on-axis Fresnel holograms, off-axis Fresnel holograms, and Fourier holograms. In this chapter we will introduce some important techniques in digital holography, including phase-shifting digital holography, low-coherence digital holography, tomographic holography, and optical scanning holography.

5.1 Phase-shifting digital holography

In optical holography, the off-axis geometry is an effective configuration for separating the zeroth-order light and the twin image. However, as indicated in Section 4.4, it is difficult to record a high-quality off-axis Fresnel hologram because of the limitation of the narrow bandwidth of the CCD. However, if the on-axis geometry is employed, it is necessary to eliminate the zeroth-order light and the twin image. Among the various available techniques, phase-shifting holography (PSH) is the most widely used technique to achieve this goal [1–3]. Figure 5.1 depicts a typical setup for PSH. The setup is nearly the same as that for conventional on-axis digital holography. The only difference is that one of the mirrors is mounted on a piezoelectric transducer (PZT) in PSH so that the optical path difference between the object light and the reference light can be adjusted. The phase difference should be precisely controlled through the PZT. Besides the use of a PZT, other phase-shifting mechanisms include the use of an electro-optic modulator [4], an acousto-optic modulator [5], a wave plate [6], and a tilted slab [7]. Phase shifting can be applied to any holographic recording geometry: on-axis holography, off-axis holography [8], or Gabor holography [9].

In PSH, we need to take multiple holograms corresponding to the various phase differences between the object light and the reference light. In general, the hologram can be expressed as

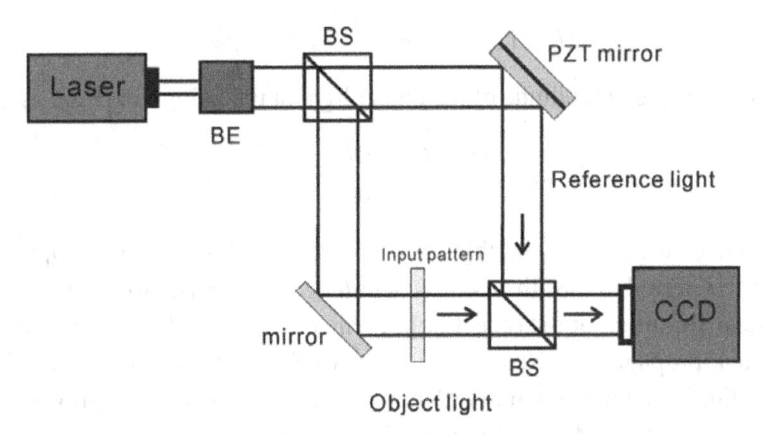

Figure 5.1 Schematic of phase-shifting holography. BE beam expander; BS beam splitter.

$$\begin{aligned} I_\delta &= |\psi_0 + \psi_r \exp(-j\delta)|^2 \\ &= |\psi_0|^2 + |\psi_r|^2 + \psi_0 \psi_r^* \exp(j\delta) + \psi_0^* \psi_r \exp(-j\delta), \end{aligned} \tag{5.1}$$

where ψ_0 and ψ_r stand for the complex amplitude of the object light and the reference light, respectively; δ stands for the phase induced by the phase shifter. Multiple holograms are acquired sequentially and, upon complete acquisition, the complex amplitude of the object light can be derived from the holograms.

5.1.1 Four-step phase-shifting holography

In four-step PSH, four holograms are acquired sequentially. The phase differences for the four holograms are $\delta = 0$, $\delta = \pi/2$, $\delta = \pi$, and $\delta = 3\pi/2$. Therefore, according to Eq. (5.1), the four holograms can be expressed as

$$\begin{aligned} I_0 &= |\psi_0|^2 + |\psi_r|^2 + \psi_0 \psi_r^* + \psi_0^* \psi_r, \\ I_{\pi/2} &= |\psi_0|^2 + |\psi_r|^2 + j\psi_0 \psi_r^* - j\psi_0^* \psi_r, \\ I_\pi &= |\psi_0|^2 + |\psi_r|^2 - \psi_0 \psi_r^* - \psi_0^* \psi_r, \\ I_{3\pi/2} &= |\psi_0|^2 + |\psi_r|^2 - j\psi_0 \psi_r^* + j\psi_0^* \psi_r. \end{aligned}$$

First, we can remove the zeroth-order light, $|\psi_0|^2 + |\psi_r|^2$, by the following subtractions:

$$I_0 - I_\pi = 2\psi_0 \psi_r^* + 2\psi_0^* \psi_r \tag{5.2a}$$

$$I_{\pi/2} - I_{3\pi/2} = 2j\psi_0 \psi_r^* - 2j\psi_0^* \psi_r. \tag{5.2b}$$

Then, the conjugate term of ψ_0, i.e., ψ_0^*, can be removed by an additional subtraction as follows:

$$(I_0 - I_\pi) - j(I_{\pi/2} - I_{3\pi/2}) = 4\psi_0\psi_r^*. \tag{5.3}$$

Finally, the complex amplitude of the object light at the hologram plane is given by

$$\psi_0 = \frac{(I_0 - I_\pi) - j(I_{\pi/2} - I_{3\pi/2})}{4\psi_r^*}. \tag{5.4}$$

In the field of digital holography, the complex amplitude of the object light is sometimes referred to as the *complex hologram* because we can retrieve the amplitude distribution of the object light in the object plane from ψ_0 by performing digital back-propagation. Note that in Eq. (5.4) the complex amplitude of the reference light must be known in order to calculate ψ_0. Otherwise, only the product of ψ_0 and ψ_r^* is retrieved. Usually, the reference light is a plane wave or a spherical wave and therefore its phase is known without any measurement.

5.1.2 Three-step phase-shifting holography

Four-step PSH can be simplified to three-step PSH, in which only three holograms are involved. The phase differences for the three holograms are $\delta = 0$, $\delta = \pi/2$, and $\delta = \pi$. After some mathematical manipulation, the complex amplitude of the object light can be found to be

$$\psi_0 = \frac{(1+j)\left(I_0 - I_{\pi/2}\right) + (j-1)(I_\pi - I_{\pi/2})}{4\psi_r^*}. \tag{5.5}$$

5.1.3 Two-step phase-shifting holography

Could the steps of phase shifting be reduced further? The answer is affirmative, and there is two-step PSH. Suppose that two holograms with a zero phase shift and a $\pi/2$ phase shift are obtained [see Eq. (5.1)]. In addition, the intensities of the object light, $|\psi_0|^2$, and the reference light, $|\psi_r|^2$, are measured. We can remove the zeroth-order light from the two holograms as follows:

$$I_0 - |\psi_0|^2 - |\psi_r|^2 = \psi_0\psi_r^* + \psi_0^*\psi_r \tag{5.6a}$$

$$I_{\pi/2} - |\psi_0|^2 - |\psi_r|^2 = j\psi_0\psi_r^* - j\psi_0^*\psi_r. \tag{5.6b}$$

As a result, the complex amplitude of the object light is found to be

$$\psi_0 = \frac{\left(I_0 - |\psi_0|^2 - |\psi_r|^2\right) - j\left(I_{\pi/2} - |\psi_0|^2 - |\psi_r|^2\right)}{2\psi_r^*}. \tag{5.7}$$

Because in this case only two holograms with a zero phase shift and a $\pi/2$ phase shift are needed, the technique is also called *quadrature-phase shifting holography*

(QPSH) [10]. In standard QPSH, the number of exposures is four (two holograms and two intensity patterns $|\psi_0|^2$ and $|\psi_r|^2$), the same as for four-step PSH. However, the recording of the intensity patterns is not sensitive to any vibration or any mechanism that causes phase variation between the object wave and the reference wave, and thus the recording of QPSH is easier than that of four-step PSH. It is also interesting to note that we do not need to measure the intensity of the object light [10–12]. To see how this happens, we first take the square of Eqs. (5.6a) and (5.6b), giving

$$I_0^2 + \left[|\psi_0|^2 + |\psi_r|^2\right]^2 - 2I_0\left[|\psi_0|^2 + |\psi_r|^2\right] = 4 \times \left[\mathrm{Re}\{\psi_0\psi_r^*\}\right]^2 \tag{5.8a}$$

$$I_{\pi/2}^2 + \left[|\psi_0|^2 + |\psi_r|^2\right]^2 - 2I_{\pi/2}\left[|\psi_0|^2 + |\psi_r|^2\right] = 4 \times \left[\mathrm{Im}\{\psi_0\psi_r^*\}\right]^2. \tag{5.8b}$$

We then add Eqs. (5.8a) and (5.8b) to obtain

$$\begin{aligned} &I_0^2 + I_{\pi/2}^2 + 2\left[|\psi_0|^2 + |\psi_r|^2\right]^2 - 2(I_0 + I_{\pi/2})\left[|\psi_0|^2 + |\psi_r|^2\right] \\ &= 4 \times \left[\mathrm{Re}\{\psi_0\psi_r^*\}\right]^2 + 4 \times \left[\mathrm{Im}\{\psi_0\psi_r^*\}\right]^2 = 4 \times |\psi_0|^2|\psi_r|^2. \end{aligned} \tag{5.9}$$

Consequently, we have a quadratic equation in $|\psi_0|^2 + |\psi_r|^2$ as follows:

$$2\left[|\psi_0|^2 + |\psi_r|^2\right]^2 - \left(4|\psi_r|^2 + 2I_0 + 2I_{\pi/2}\right)\left[|\psi_0|^2 + |\psi_r|^2\right] + \left(I_0^2 + I_{\pi/2}^2 + 4|\psi_r|^4\right) = 0. \tag{5.10}$$

The solution to Eq. (5.10) is

$$|\psi_0|^2 + |\psi_r|^2 = \frac{2|\psi_r|^2 + I_0 + I_{\pi/2}}{2} \pm \frac{\sqrt{(2|\psi_r|^2 + I_0 + I_{\pi/2})^2 - 2(I_0^2 + I_{\pi/2}^2 + 4|\psi_r|^4)}}{2}. \tag{5.11}$$

As a result, the intensity of the object light $|\psi_0|^2$ can be calculated using Eq. (5.11), provided that I_0, $I_{\pi/2}$, and $|\psi_r|^2$ are known.

Note that in Eq. (5.11), only one of the +/– signs corresponds to a correct solution. The selection of the sign depends on the intensity ratio of the object light and the reference light, and the phase difference between the object light and the reference light, i.e., $\arg\{\psi_0\psi_r^*\}$. To see this dependence, we first choose the minus sign and calculate the intensity using Eq. (5.11). The relative error of the calculated intensity of the object light is shown in Fig. 5.2, and is defined as

$$\text{error} = \left|1 - \frac{I_0 + I_{\pi/2} - \sqrt{(2|\psi_r|^2 + I_0 + I_{\pi/2})^2 - 2(I_0^2 + I_{\pi/2}^2 + 4|\psi_r|^4)}}{2|\psi_0|^2}\right|. \tag{5.12}$$

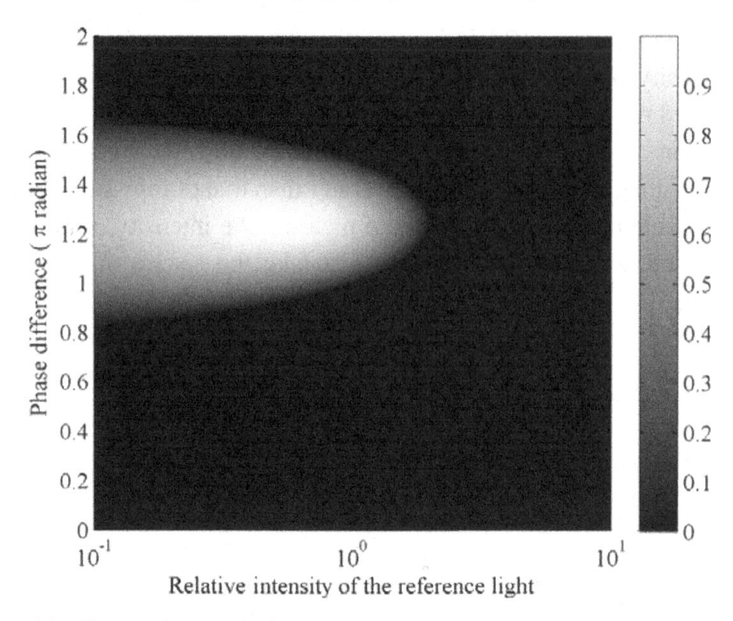

Figure 5.2 Error of the calculated intensity of the object light as a function of both the phase difference, $\arg\{\psi_0\psi_r^*\}$, and the relative intensity, $|\psi_r|^2/|\psi_0|^2$. See Table 5.1 for the MATLAB code.

In Fig. 5.2, the gray-level represents the amount of error. We can see that the error in the range of 2π is small only when the relative intensity, defined as $|\psi_r|^2/|\psi_0|^2$, is larger than 2. Accordingly, to ensure that the calculated intensity of the object light at any pixel is correct, the intensity of the reference light must be at least two times larger than the maximum intensity of the object light. This demand limits the dynamic range of the acquired digital holograms, and thus limits the signal-to-noise ratio of the reconstructed images. When the plus sign in Eq. (5.11) is chosen, results indicate that not only the relative intensity but also $\arg\{\psi_0\psi_r^*\}$ must be limited to a small range to ensure a small error. So the minus sign in Eq. (5.11) is always selected.

5.1.4 Phase step and phase error

In the above analysis, the phase step is always $\pi/2$ for four-step PSH, three-step PSH, and two-step PSH. However, the complex object field can be recorded for any phase step ranging between 0 and π. Here we take two-step PSH as an example. For an arbitrary phase step δ, the two holograms acquired are

$$I_1 = |\psi_0|^2 + |\psi_r|^2 + \psi_0\psi_r^* + \psi_0^*\psi_r, \tag{5.13a}$$

$$I_2 = |\psi_0|^2 + |\psi_r|^2 + \psi_0\psi_r^* \exp(j\delta) + \psi_0^*\psi_r \exp(-j\delta). \tag{5.13b}$$

Table 5.1 *MATLAB code for calculating the error of QPSH, see Fig. 5.2*

```
clear all; close all;
ii=logspace(-1,1,1024);
theta=linspace(0,2,1024);
[I,T]=meshgrid(ii,theta);
I1=1+I+2*(I.^0.5).*cos(T*pi); %first hologram
I2=1+I+2*(I.^0.5).*sin(T*pi); %second hologram
D=8*I.*I2-(I1-I2-2*I).^2;
Io=(I1+I2-D.^0.5)./2;
err=abs(Io-1); % The error
figure; mesh(I,T,err);
view(2)
set(gca, 'XScale', 'log') ;
xlabel('Relative Intensity of the reference light');
ylabel('Phase difference (\pi radian)');
colorbar
axis tight
```

Equations (5.13a) and (5.13b) can be rearranged to become

$$\psi_o\psi_r^* + \psi_o^*\psi_r = I_1 - |\psi_o|^2 - |\psi_r|^2, \tag{5.14a}$$

$$\psi_o\psi_r^*\exp(j\delta) + \psi_o^*\psi_r\ \exp(-j\delta) = I_2 - |\psi_o|^2 - |\psi_r|^2. \tag{5.14b}$$

We now subtract Eq. (5.14b) from Eq. (5.14a) after multiplying Eq. (5.14a) by $\exp(-j\delta)$. The result of the subtraction is

$$-2j\psi_0\psi_r^*\sin(\delta) = \left(I_1 - |\psi_0|^2 - |\psi_r|^2\right)\exp(-j\delta) - \left(I_2 - |\psi_0|^2 - |\psi_r|^2\right). \tag{5.15}$$

Consequently, the complex amplitude of the object wave can be retrieved by

$$\psi_0 = \frac{\left(I_1 - |\psi_0|^2 - |\psi_r|^2\right)\exp(-j\delta) - \left(I_2 - |\psi_0|^2 - |\psi_r|^2\right)}{-2j\psi_r^*\ \sin(\delta)}. \tag{5.16}$$

According to Eq. (5.16), the phase step δ can be arbitrary in the range $0 < \delta < \pi$ or $-\pi < \delta < 0$.

Equation (5.16) shows that the phase step can be arbitrary in a wide range. However, in practice, tolerance of the phase step in phase-shifting holography is a problem. Tolerance may be due to the inaccuracy of the phase shifter, air turbulence, or vibration. It can be easily shown that for a fixed tolerance of the phase step, the minimum error of the retrieved object field can be obtained when the $\pi/2$ (or $-\pi/2$) phase step is applied.

There are some techniques for eliminating the phase error or for measuring the actual phase step used so that a high-quality reconstructed image can be obtained. For example, if the phase applied in retrieving the object field deviates from the phase step applied in the experiment, the twin image cannot be removed completely. Thus the correct phase step can be determined to be the phase at which the power of the residual twin image is minimized. This calibration method can be easily implemented for off-axis holography [8], but not for on-axis holography as the twin image and the zeroth-order beam overlap. Hence we cannot evaluate the phase by monitoring the twin image.

For on-axis holography, one can use other constraints as a measurement of the residual power of the twin image. For example, the amplitude should be uniform across the object plane if the object is a purely phase object. When the phase applied in retrieving the object light is correct, the amplitude fluctuation at the object plane should be minimized [13, 14]. Other algorithms have been proposed to directly estimate the sequential phase step process between the holograms [15, 16].

5.1.5 Parallel phase-shifting holography

In conventional PSH, multiple exposures are necessary for acquiring the different holograms and the intensity patterns if the numbers of holograms are to be reduced. It appears difficult to acquire a single-exposure hologram which will contain the different phase-shifted holograms. Nevertheless, parallel phase-shifting holography (PPSH) [15, 16] can acquire such a single-exposure hologram by "partitioning" the resolution of a CCD. The schematic of PPSH is shown in Fig. 5.3. In PPSH, the phase shifter is replaced by a phase mask of an array of "super pixels." Each super pixel consists of four pixels with phase retardation of 0, $\pi/2$, π, and $3\pi/2$. The reference light passing through the phase mask is then imaged on the CCD so that the image of the phase mask and the CCD chip are matched pixel by pixel.

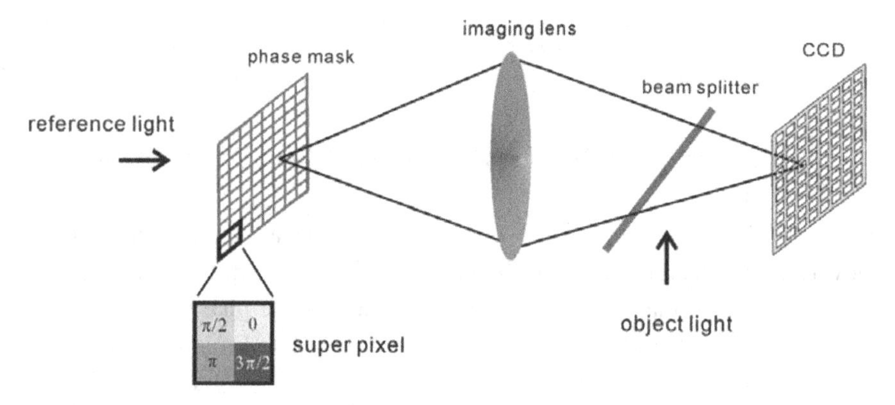

Figure 5.3 Schematic of parallel phase-shifting holography.

In addition, the object light is also directed to the CCD by a beam splitter. As a result, interferences on the four pixels of a super pixel correspond to four different phase steps. Subsequently, we can calculate the complex amplitude of the object wave using Eq. (5.4). The pixel size of the retrieved object light thus has the same size as a super pixel, but is four times larger than the pixel size of the CCD.

The pixel-by-pixel imaging in Fig. 5.3 demands a high-precision alignment and is sensitive to environmental interference. An improvement is made by attaching the phase mask directly against the CCD chip [17–19]. Alternatively, a random phase mask can be used instead of a deterministic phase mask [20]. However, several pre-measurements must be performed in order to calibrate the phase of the reference light among super pixels.

Example 5.1: Simulation of four-step PSH

In this example we simulate the recording and reconstruction of four-step phase-shifting holography (PSH). Since PSH can be conducted using the geometry of on-axis holography, we adopt the simulation setup of on-axis holography (see Example 4.7). The only difference is that we perform the recording four times, corresponding to the phase shifts 0, $\pi/2$, π, and $3\pi/2$. The four holograms are shown in Fig. 5.4. As the holograms are recorded, we can find the complex holograms using Eq. (5.4) for four-step PSH. The complex hologram can be reconstructed using standard backward propagation. The reconstructed image of the complex hologram is shown in

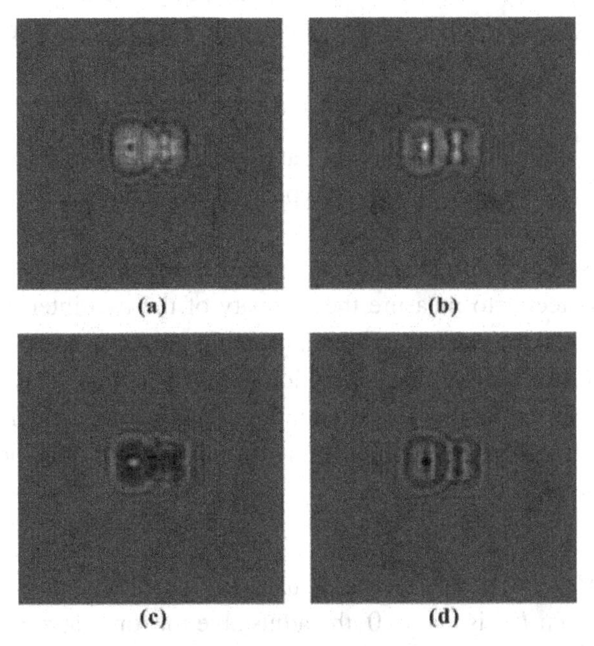

Figure 5.4 Four holograms for (a) $\delta = 0$, (b) $\delta = \pi/2$, (c) $\delta = \pi$, and (d) $\delta = 3\pi/2$.

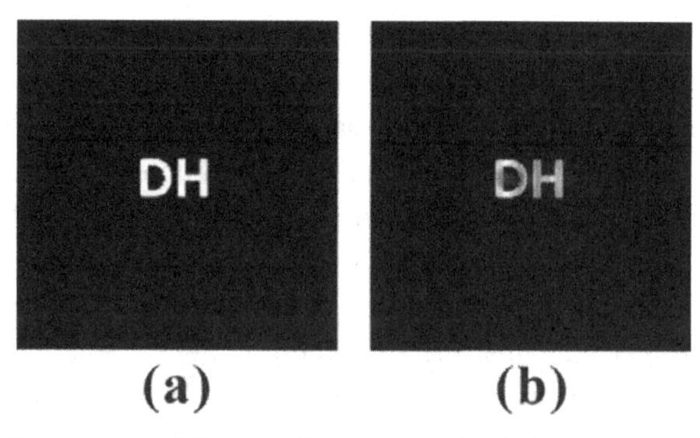

Figure 5.5 Reconstructed image of (a) a complex hologram containing correct phase steps, and (b) a complex hologram containing phase errors.

Fig. 5.5(a), which is identical to the original pattern [Fig. 4.18(a)]. There is no noticeable zeroth-order light or twin image among the reconstructed image as compared to that in conventional on-axis holographic reconstruction [Fig. 4.18(d)]. In another simulation, we set the phase steps to be 0, 0.25π, 0.5π, and 0.75π. The complex hologram is still obtained using Eq. (5.4). But as a result of the phase errors, the complex hologram is not correct. The corresponding reconstructed image is shown in Fig. 5.5(b). We can see that the image quality is a little worse than that shown in Fig. 5.5(a). In any case, there is no zeroth-order light present in the reconstruction. This is because in producing the complex hologram, the zeroth-order light can be removed even though the phase steps are not correct [Eq. (5.14)]. The MATLAB code for this example is listed in Table 5.2 as a reference.

5.2 Low-coherence digital holography

Low-coherence digital holography is based on the principle of low-coherence interferometry. Figure 5.6 shows a typical Michelson interferometer used to characterize the coherence of a source.

Lens L_1 is used to collimate the source and lens L_2 is used to collect the light onto the photodetector to measure the intensity of the two interfering light fields reflected from the fixed mirror and the adjustable mirror, where the movement of the adjustable mirror is along the z-direction to realize a time delay τ. For example, if the light source is a plane wave after collimation by lens L_1, $A(t) = A_0 e^{j\omega_0 t}$ and $B(t) = A_0 e^{j\omega_0 (t+\tau)}$ according to Eqs. (2.41a) and (2.41b), and the interferogram then becomes, using Eq. (2.45),

$$I(\tau) = 2I_0[1 + \cos(\omega_0 \tau)], \tag{5.17}$$

where $I(0) = 4I_0 = 4A_0^2$ as $\gamma(\tau) = \exp(j\omega_0\tau)$ and $|\gamma(\tau)| = 1$, previously discussed in Chapter 2. When $I(\tau)$ is at $\tau = 0$, the adjustable mirror is at $z = 0$ (see Fig. 5.6) and there is no delay of the two interfering plane waves. Figure 5.7(a) shows $I(\tau) - 2I_0$

Table 5.2 *MATLAB code for simulation of four-step PSH, see Example 5.1; the code is based on simulation of an on-axis hologram (Example 4.7)*

```
clear all; close all;
I=imread('DH256.bmp','bmp'); %256×256 pixels 8bit image
I=double(I);
% parameter setup
M=256;
deltax=0.001; % pixel pitch 0.001 cm (10 um)
w=633*10^-8; % wavelength 633 nm
z=25; % 25 cm, propagation distance
delta=pi/2; % phase step (change it to show part b)

%Step 1: simulation of propagation using the ASM
r=1:M;
c=1:M;
[C, R]=meshgrid(c, r);
A0=fftshift(ifft2(fftshift(I)));
deltaf=1/M/deltax;
p=exp(-2i*pi*z.*((1/w)^2-((R-M/2-1).*deltaf).^2-...
    ((C-M/2-1).*deltaf).^2).^0.5);
Az=A0.*p;
EO=fftshift(fft2(fftshift(Az)));

%Step 2: Interference at the hologram plane
AV=(min(min(abs(EO)))+max(max(abs(EO))));
% the amplitude of reference light
% Recording of Four phase-shifting holograms
I0=(EO+AV).*conj(EO+AV);
I1=(EO+AV*exp(-1j*delta)).*conj(EO+AV*exp(-1j*delta));
I2=(EO+AV*exp(-2j*delta)).*conj(EO+AV*exp(-2j*delta));
I3=(EO+AV*exp(-3j*delta)).*conj(EO+AV*exp(-3j*delta));
MAX=max(max([I0, I1, I2, I3]));
figure(1); imshow(I);
title('Original object')
axis off
figure(2)
subplot(2,2,1)
imshow(I0/MAX);
axis off
title('hologram 1')
subplot(2,2,2)
imshow(I1/MAX);
axis off
title('hologram 2')
subplot(2,2,3)
imshow(I2/MAX);
axis off
title('hologram 3')
subplot(2,2,4)
```

Table 5.2 (*cont.*)

```
imshow(I3/MAX);
axis off
title('hologram 4')

%Step 3: Reconstruction
CH=(I0-I2)-1j*(I1-I3); % the complex hologram (4-step PSH)
A1=fftshift(ifft2(fftshift(CH)));
Az1=A1.*conj(p);
EI=fftshift(fft2(fftshift(Az1)));
EI=(EI.*conj(EI));
EI=EI/max(max(EI));
figure(3);
imshow(EI);
title('Reconstructed image of 4-step PSH')
axis off
```

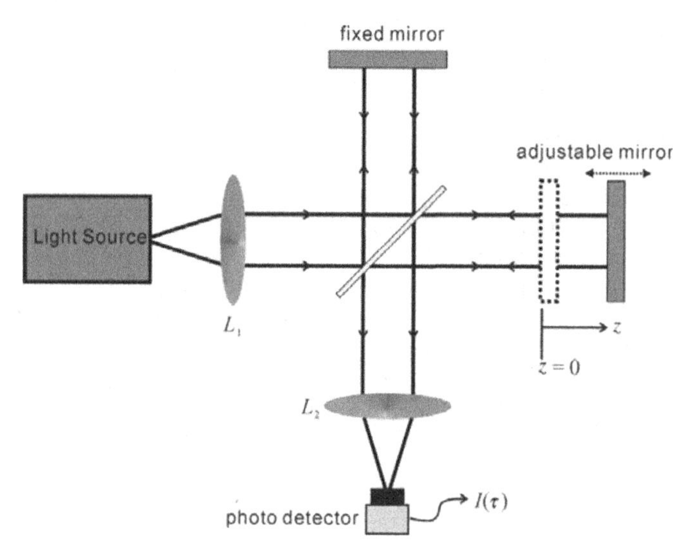

Figure 5.6 Michelson interferometer used to characterize the coherence of a source

versus τ, i.e., the bias has been subtracted from $I(\tau)$ and only the intensity variation is plotted, for a plane wave source. In this case, we have complete coherence.

For a partially coherent source having a sinc-squared type power spectrum [see Eq. (2.63)], we have

$$I(\tau) = 2I_0\left[1 + \Lambda\left(\frac{\tau}{\tau_0}\right)\cos\left(\omega_0\tau\right)\right] \tag{5.18}$$

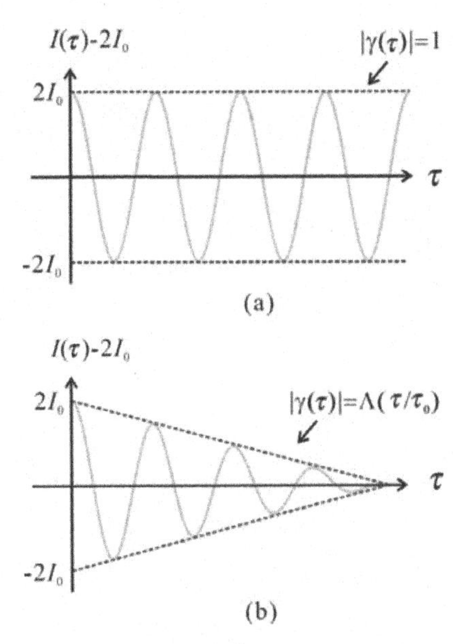

Figure 5.7 $I(\tau) - 2I_0$ versus τ for (a) a complete coherent source, and (b) a source having a sinc-squared type power spectrum.

according to Eq. (2.49). Figure 5.7(b) shows the intensity variation as a function of τ, a phenomenon in which an interference pattern forms only when the optical path difference between the two incident fields is less than the coherence length of the light source, $\ell_c = c \times \tau_0$. Low-coherence holography uses this phenomenon to allow for interference between the object wave and the reference wave. The thickness of the layer of the object that can be holographically recorded is about half of the coherence length owing to the reflection setup. Since ℓ_c can be made short by using an appropriate source, the thickness of the layer of a thick object being recorded can be made small. The capability of imaging only a thin section optically without any out-of-focus haze, i.e., noise coming from sections other than the section being imaged, is called *optical sectioning* [21–24]. We shall adopt the holographic microscope investigated by Lin *et al.* [21] to illustrate the principle of low-coherence holography.

The principle of low-coherence holography is easily understood now that we have discussed low-coherence interferometry. The experimental setup of a low-coherence phase-shifting digital holographic microscope for measuring sectional images is shown in Fig. 5.8. It consists of a modified Michelson interferometer with a PZT-driven mirror in the reference arm and a high resolution CCD sensor (pixel number, 1280 × 1024; pixel size, 5.2 μm × 5.2 μm). The motorized linear stage is moved longitudinally step by step to different depths of the sample to capture holograms at different layers. Then, the PZT-driven

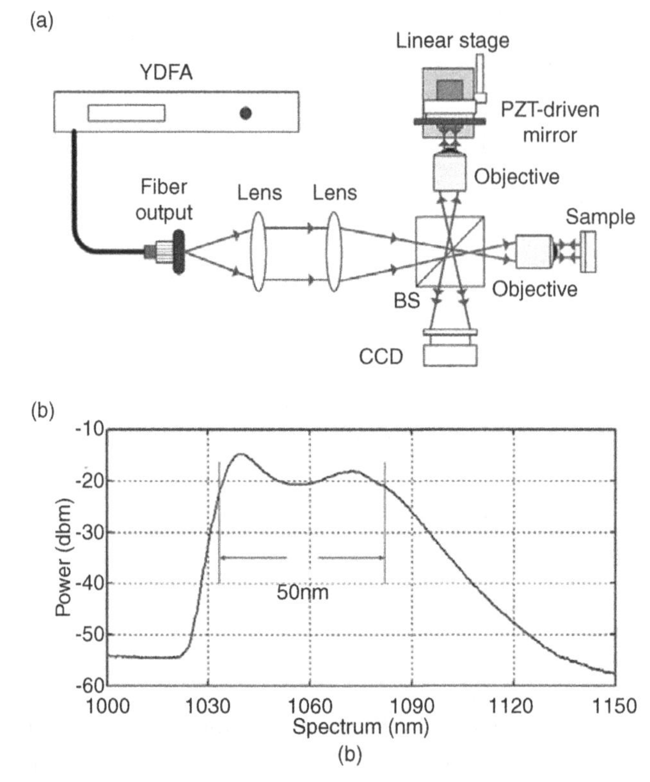

Figure 5.8 (a) Experimental setup of a low-coherence phase-shifting digital holographic microscope and (b) spectral shape of the YDFA, with FWHM of 50 nm. Reprinted from Ref. [21], with permission, © OSA.

mirror is used to achieve the phase-shifting algorithm for hologram recording with four-step phase intervals of 0, $\pi/2$, π, and $3\pi/2$ for each layer. The objective is 10× with a *NA* of 0.25, corresponding to a transverse resolution of about 3 μm. The low-coherence light source is a ytterbium-doped fiber amplifier (YDFA), whose spectral profile is shown in Fig. 5.8(b). The center wavelength λ_0 is about 1060 nm, and the spectral bandwidth $\Delta\lambda$ is 50 nm at full width at half-maximum (FWHM). The coherence length, $\lambda_0^2/\Delta\lambda$, determines the thickness of the sectional layer, which is estimated to be about 15 μm. Similar to Eq. (2.49), the intensity of the four phase-shifted interference fringes presented as digital holograms is generated by the interference of the object wave $E_0 = |E_0|e^{-j\phi_0}$ and reference wave $E_R = |E_R|e^{-j\phi_R}$ in the CCD plane as follows:

$$I_{(k-1)\pi/2} \propto |E_0|^2 + |E_R|^2 + 2|E_0||E_R||\gamma(\tau)|\cos[\phi_0 - \phi_R + \omega_0\tau], \tag{5.19}$$

where $k = 1,2,3,4$ is for the different phase shifts in four-step phase shifting holography. We have borrowed the result from Eq. (2.45) to fit into our current holographic recording situation. In Eq. (5.19), $\tau = 2(l_0 - l_{R_k})/c$, where l_0 denotes the distance between the object and the center of the beam splitter (BS) and l_{R_k} is the distance between the reference mirror and the center of the beam splitter, and by changing its distance for different k, we can realize the different phase shifts between the object wave and the reference wave. Now in Eq. (5.19), ϕ_R can be grouped with $\omega_0\tau$ because l_{R_k} is a variable, hence, we can write

$$\omega_0\tau - \phi_R = 2k_0(l_0 - l_{R_k}), \tag{5.20}$$

where $k_0 = \omega_0/c$. For $\gamma(\tau)$, we use the model of the spectral shape of the low-coherence light source as a rectangular shape and hence $|\gamma(\tau)| = \mathrm{sinc}(\Delta\omega\tau/2\pi)$, according to Eq. (2.65b). Using Eq. (2.69), the argument $\Delta\omega\tau/2\pi$ can be expressed as follows:

$$\frac{\Delta\omega\tau}{2\pi} = \frac{\Delta\omega}{2\pi}\frac{2(l_0 - l_{R_k})}{c} = \frac{\Delta\lambda}{\lambda_0^2}2(l_0 - l_{R_k}). \tag{5.21}$$

Since $2(l_0 - l_{R_k})$ can be changed to obtain different phase shifts for path lengths of 0, $\lambda_0/4$, $\lambda_0/2$, $3\lambda_0/4$, which correspond to phase shifts of $(k - 1)\pi/2$, we have

$$k_0 2(l_0 - l_{R_k}) = (k-1)\pi/2, \tag{5.22}$$

and Eq. (5.21) can be re-written as

$$\frac{\Delta\omega\tau}{2\pi} = (k-1)\frac{\Delta\lambda}{4\lambda_0}. \tag{5.23}$$

Using Eqs. (5.20)–(5.23), together with the functional form of $|\gamma(\tau)|$ we can re-write Eq. (5.19) as

$$I_{(k-1)\pi/2} \propto |E_0|^2 + |E_R|^2 + 2|E_0||E_R|\left|\mathrm{sinc}\left[(k-1)\frac{\Delta\lambda}{4\lambda_0}\right]\right|\cos\left[\phi_0 + (k-1)\pi/2\right]. \tag{5.24}$$

By using the phase-shifting procedure, i.e., according to Eq. (5.4) and using Eq. (5.24), we can extract the complex field $E_0 = |E_0|e^{-j\phi_0}$ with the phase and the amplitude, respectively, as follows:

$$\phi_0 = \arctan\left\{\frac{I_{3\pi/2} - I_{\pi/2}}{I_0 - I_\pi}\frac{\left[\mathrm{sinc}\left(\frac{\Delta\lambda}{2\lambda_0}\right) + 1\right]}{\left[\mathrm{sinc}\left(\frac{3\Delta\lambda}{4\lambda_0}\right) + \mathrm{sinc}\left(\frac{\Delta\lambda}{4\lambda_0}\right)\right]}\right\}, \tag{5.25}$$

and

$$|E_0|=\frac{\sqrt{\left\{(I_{3\pi/2}-I_{\pi/2})\left[\text{sinc}\left(\frac{\Delta\lambda}{2\lambda_0}\right)+1\right]\right\}^2+\left\{(I_0-I_\pi)\left[\text{sinc}\left(\frac{3\Delta\lambda}{4\lambda_0}\right)+\text{sinc}\left(\frac{\Delta\lambda}{4\lambda_0}\right)\right]\right\}^2}}{2\left[\text{sinc}\left(\frac{\Delta\lambda}{2\lambda_0}\right)+1\right]\left[\text{sinc}\left(\frac{3\Delta\lambda}{4\lambda_0}\right)+\text{sinc}\left(\frac{\Delta\lambda}{4\lambda_0}\right)\right]} \tag{5.26}$$

by assuming $E_R = 1$.

Figure 5.9 shows a tranquilized zebra fish. Holograms were captured using the CCD at 2.5 frames/second. Figure 5.10(a) shows a portion of the tail of the zebra fish where holograms were captured, and Fig. 5.10(b) – (d) show three optical sections. Clearly, in Fig. 5.10(d), we see the outline of the spine at that section. Figure 5.11 shows the phase plot at the section $z = 60$ μm, where the spine is clearly outlined.

Current low-coherence holographic recording systems have a sectioning capability of about 10–20 μm as is the case shown in the above results. When using an ultra-broadband source, such as a supercontinuum of spectral bandwidth of about 500 nm, a very short coherence length of the order of 1 μm is possible.

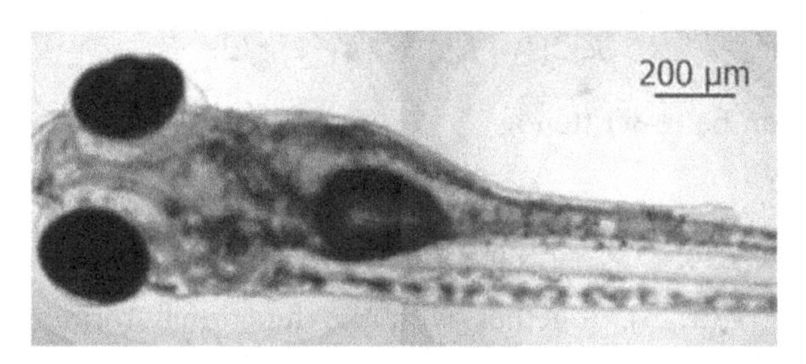

Figure 5.9 Image of a zebra fish. Reprinted from Ref. [21], with permission, © OSA.

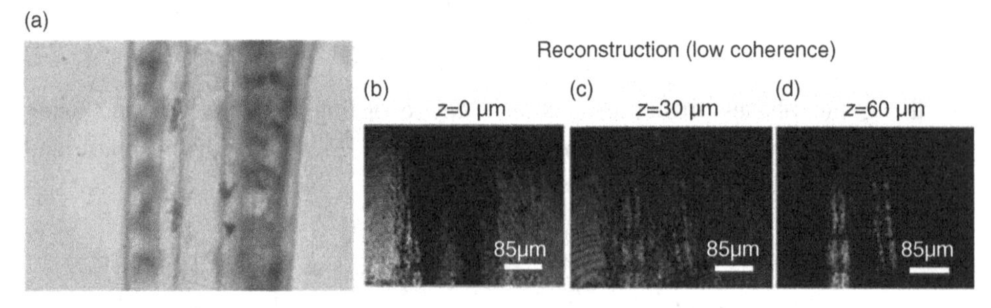

Figure 5.10 (a) Portion of the tail to be recorded holographically. (b)–(d) Optical sections at three different sections. Reprinted from Ref. [21], with permission, © OSA.

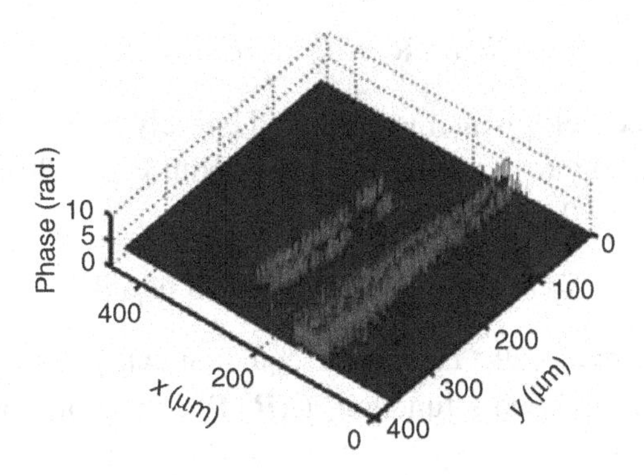

Figure 5.11 Phase plot at the section $z = 60$ μm shown in Fig. 5.10(d). Reprinted from Ref. [21], with permission, © OSA.

5.3 Diffraction tomographic holography

Tomography refers to imaging by sectioning. Low-coherence digital holography discussed in the previous section is actually an example of tomographic holography. In this section, we discuss diffraction tomographic holography, where the three-dimensional structure of a semi-transparent object such as a biological cell can be obtained from holographic data. To be precise, the objective of diffraction tomographic holography is to image the refractive index, $n(\mathbf{R})$, which serves as an important intrinsic contrast agent, in order to visualize biological cells. To understand diffraction tomographic holography, let us first discuss the principle of the Fourier diffraction theorem in tomography. We start with the Helmholtz equation for ψ_p in an inhomogeneous medium characterized by $n(\mathbf{R})$,

$$\nabla^2\psi_p + k_0^2 n^2(\mathbf{R})\psi_p = 0, \tag{5.27}$$

where for convenience we use the symbol $\mathbf{R}$ to denote x, y, and z. We can re-write Eq. (5.27) as

$$\nabla^2\psi_p + k_0^2\psi_p + k_0^2[n^2(\mathbf{R}) - 1]\psi_p = 0,$$

or

$$\nabla^2\psi_p(\mathbf{R}) + k_0^2\psi_p(\mathbf{R}) = o(\mathbf{R})\psi_p(\mathbf{R}), \tag{5.28}$$

where $o(\mathbf{R}) = -k_0^2[n^2(\mathbf{R}) - 1]$ is known as the object function, which was originally called the scattering potential of the object as it characterizes the scattering object [25]. We now let the total field ψ_p be

$$\psi_p(\mathbf{R}) = \psi^i(\mathbf{R}) + \psi^s(\mathbf{R}), \tag{5.29}$$

where $\psi^i(\mathbf{R})$ is the incident field, which is the field present without any inhomogeneities in the medium, and therefore satisfies the following equation:

$$\nabla^2\psi^i(\mathbf{R}) + k_0^2\psi^i(\mathbf{R}) = 0. \tag{5.30}$$

$\psi^s(\mathbf{R})$ is the scattered field, which is attributed entirely to the inhomogeneities. By substituting Eq. (5.29) into Eq. (5.28), together with Eq. (5.30), we can obtain the following equation for the scattered field:

$$\nabla^2\psi^s(\mathbf{R}) + k_0^2\psi^s(\mathbf{R}) = o(\mathbf{R})\psi_p(\mathbf{R}). \tag{5.31}$$

A solution to the above scalar Helmholtz equation can be written in terms of the Green's function. The Green's function, $G(\mathbf{R}, \mathbf{R}')$, is a solution of the following equation:

$$\nabla^2 G(\mathbf{R}, \mathbf{R}') + k_0^2 G(\mathbf{R}, \mathbf{R}') = \delta(\mathbf{R} - \mathbf{R}') \tag{5.32}$$

with

$$G(\mathbf{R}, \mathbf{R}') = -\frac{e^{-jk_0|\mathbf{R}-\mathbf{R}'|}}{4\pi|\mathbf{R}-\mathbf{R}'|} = G(|\mathbf{R}-\mathbf{R}'|).$$

Now, the source term in Eq. (5.31) can be written as a collection of impulses weighted by $o(\mathbf{R}')\ \psi_p(\mathbf{R}')$ and shifted by $\mathbf{R}'$:

$$o(\mathbf{R})\psi_p(\mathbf{R}) = \int o(\mathbf{R}')\psi_p(\mathbf{R}')\delta(\mathbf{R}-\mathbf{R}')d^3\mathbf{R}'. \tag{5.33}$$

Therefore, the solution to Eq. (5.31) can be written as

$$\psi^s(\mathbf{R}) = \int G(\mathbf{R}-\mathbf{R}')o(\mathbf{R}')\psi_p(\mathbf{R}')d^3\mathbf{R}'. \tag{5.34}$$

Note that in Eq. (5.34), we still have the unknown total field, $\psi_p(\mathbf{R}')$. To find the solution to the scattered field, we make use of one of the simplest approximations called the *first Born approximation.* Other approximations such as the first Rytov approximation have also been used to estimate the solution to the scattered field [26]. The first Born approximation states that the scattered field is much weaker than the incident field, i.e.,

$$\psi^s(\mathbf{R}) \ll \psi^i(\mathbf{R}). \tag{5.35}$$

Under this approximation, we can replace $\psi_p(\mathbf{R})$ by $\psi^i(\mathbf{R})$ in Eq. (5.34) to obtain the Born approximation of the scattered field, given by

$$\psi^s(\mathbf{R}) = \int G(\mathbf{R}-\mathbf{R}')o(\mathbf{R}')\psi^i(\mathbf{R}')d^3\mathbf{R}'. \tag{5.36}$$

Based on the above result, we will now derive what is known as the Fourier diffraction theorem in tomography. The theorem will relate the Fourier transform of the scattered field to the three-dimensional Fourier transform of the object function. The Green's function is a spherical wave and we now introduce Weyl's plane wave expansion of a spherical wave [27]:

$$G(|\mathbf{R}-\mathbf{R}'|) = -\frac{e^{-jk_0|\mathbf{R}-\mathbf{R}'|}}{4\pi|\mathbf{R}-\mathbf{R}'|} = \frac{j}{8\pi^2}\iint_{-\infty}^{\infty} \frac{e^{-j[k_x(x-x')+k_y(y-y')+k_z(z-z')]}}{\sqrt{k_0^2-k_x^2-k_y^2}} dk_x dk_y, \quad (5.37)$$

where $k_z = \sqrt{k_0^2-k_x^2-k_y^2}$ under the restriction that $k_x^2+k_y^2 < k_0^2$ so that evanescent waves are ignored. Now assuming that the incident field is a plane wave propagating along the z-direction, i.e., $\psi^i(\mathbf{R}) = e^{-jk_0 z}$, and measuring the scattered field at $z = l$ as shown in Fig. 5.12, Eq. (5.36) becomes

$$\psi^s(x,y;z=l) = \iiint_{-\infty}^{\infty}\left[\frac{j}{8\pi^2}\iint_{-\infty}^{\infty}\frac{e^{-j[k_x(x-x')+k_y(y-y')+k_z(l-z')]}}{\sqrt{k_0^2-k_x^2-k_y^2}} dk_x\, dk_y\; o(x',y',z')e^{-jk_0 z'}\right] \times dx'dy'dz'. \quad (5.38)$$

We evaluate the primed variables first, giving the following three-dimensional Fourier transform of $o(x,y,z)e^{-jk_0 z}$:

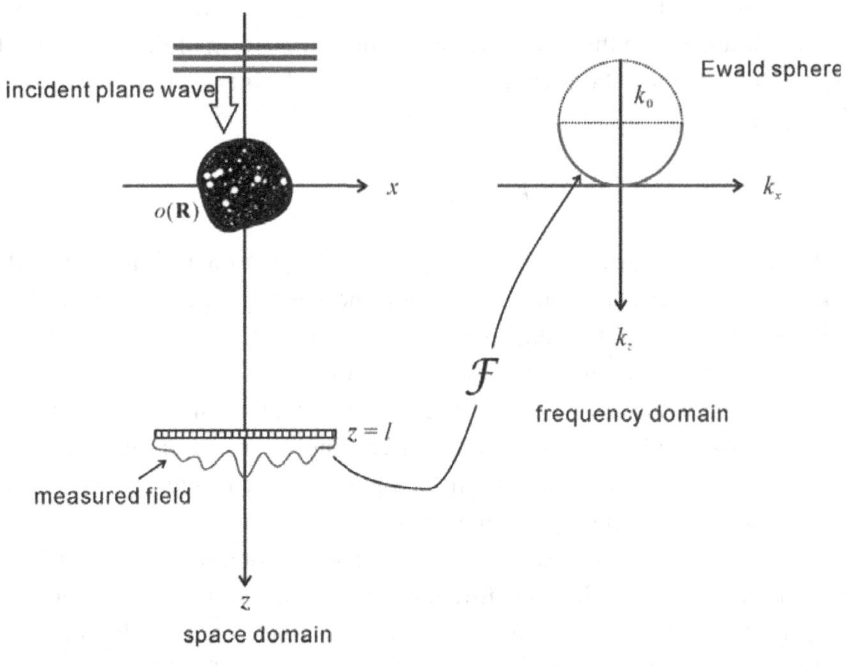

Figure 5.12 The two-dimensional scattered field spectrum along the plane $z = l$ is related to the Fourier transform of the object on the surface of the Ewald sphere.

$$\iiint_{-\infty}^{\infty} e^{jk_x x' + jk_y y' + jk_z z'} o(x', y', z') e^{-jk_0 z'}\, dx' dy' dz' = \mathcal{F}_{3D}\{o(x, y, z,)e^{-jk_0 z}\}$$

$$= O(k_x, k_y, k_z - k_0). \tag{5.39}$$

Putting the above result into Eq. (5.38), we have

$$\psi^s(x, y; z = l) = \frac{j}{8\pi^2} \iint_{-\infty}^{\infty} \frac{O(k_x, k_y, k_z - k_0)}{\sqrt{k_0^2 - k_x^2 - k_y^2}} e^{-jk_x x - jk_y y - jk_z l} dk_x dk_y$$

$$= \frac{j}{2} \mathcal{F}^{-1} \left\{ \frac{O(k_x, k_y, k_z - k_0)}{\sqrt{k_0^2 - k_x^2 - k_y^2}} e^{-j\sqrt{k_0^2 - k_x^2 - k_y^2}\, l} \right\},$$

which can be recognized as a two-dimensional Fourier transform as shown in the last step of the equation. By taking the two-dimensional Fourier transform of the above equation, we have

$$-2j\mathcal{F}\{\psi^s(x, y; z = l)\} e^{j\sqrt{k_0^2 - k_x^2 - k_y^2}\, l} \sqrt{k_0^2 - k_x^2 - k_y^2} = O(k_x, k_y, k_z - k_0), \tag{5.40}$$

where we have used $k_z = \sqrt{k_0^2 - k_x^2 - k_y^2}$. Equation (5.40) is the Fourier diffraction theorem as it relates the three-dimensional Fourier transform of the object function to the two-dimensional Fourier transform of the field measured at the recording device. Note that

$$k_z = \sqrt{k_0^2 - k_x^2 - k_y^2} \tag{5.41}$$

is the *Ewald equation*, which says that the frequency spectrum is limited to a sphere, $k_x^2 + k_y^2 + k_z^2 = k_0^2$ in the frequency domain, the so-called *Ewald sphere*. Hence, Eq. (5.40) states that the two-dimensional scattered field spectrum along the plane $z = l$ is related to the Fourier transform of the object on the surface of the Ewald sphere. The situation is shown in Fig. 5.12. For brevity, we only illustrate along two directions x and z and therefore k_x and k_z, accordingly. The solid line of the Ewald sphere can be considered transmission tomography, whereas the dotted line indicates reflection tomography where the measured field is on the plane $z = -l$.

The power of diffraction tomography is that on rotating the object by 360 degrees, each rotation will map different regions of the three-dimensional frequency spectrum of the object function. After a complete mapping by rotating, we can take the inverse three-dimensional Fourier transform of the object function to get the three-dimensional complex refractive index. Figure 5.13 summarizes the

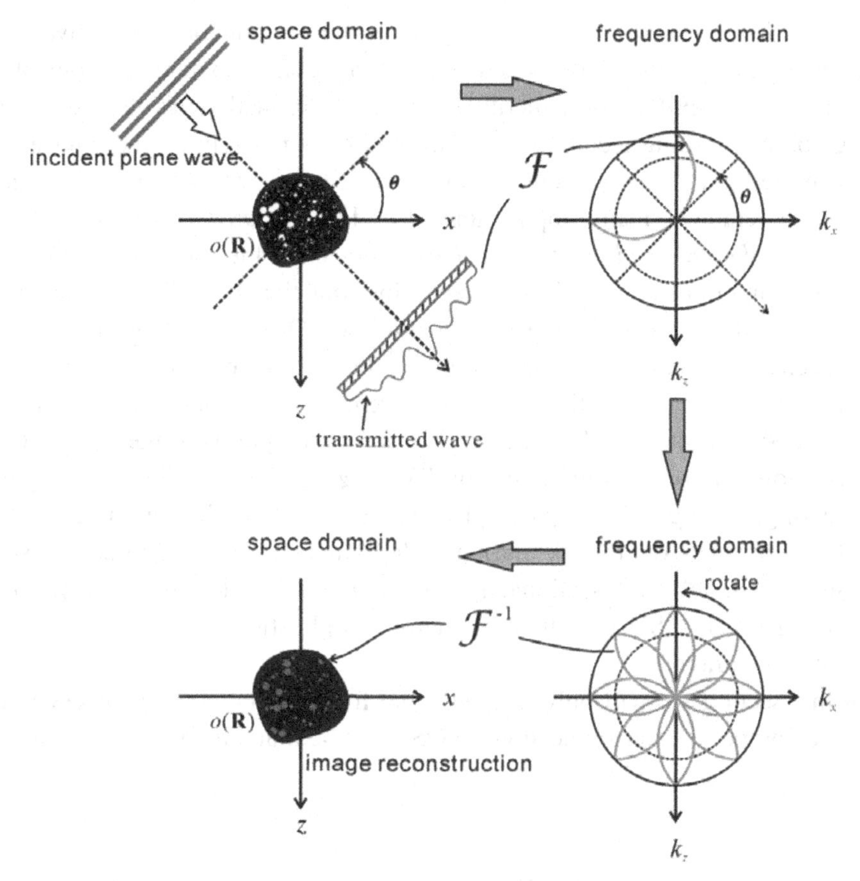

Figure 5.13 Summary of Fourier diffraction tomography.

process of diffraction tomography, where the three-dimensional spectrum is filled with eight rotations as an example with the object being rotated on the x–z plane.

In diffraction holographic tomography, we just need to record the complex scattered field $\psi^s(x, y; z = l)$ holographically for many incident angles and then perform the Fourier diffraction theorem for reconstruction of the three-dimensional object.

5.4 Optical scanning holography

Optical scanning holography (OSH) is a unique digital holographic recording technique proposed by Poon and Korpel [28, 29]. In OSH, a structured optical beam is produced by interference and the object target to be measured is raster scanned using the interference pattern. Meanwhile, the holographic data are acquired pixel by pixel. The holographic data can be directly displayed on a real-time electronic device, such as a cathode-ray tube (CRT) monitor. Alternatively, the holographic data can be digitized and stored as a digital hologram in a computer.

OSH has several unique merits in comparison with conventional digital holographic recording techniques. First, OSH does not rely on the interference between the light scattered from an object and a reference light. So OSH can be applied to remote sensing applications. In addition, OSH can be operated in either the *coherent mode* or the *incoherent mode*. In the incoherent mode, the phase information of the object target is lost but the three-dimensional information is still recorded. So the incoherent-mode OSH can be used to acquire fluorescence holograms [30–33] in the field of bio-sensing. Moreover, the acquired hologram as well as the interference pattern applied in OSH can be modified by adjusting the functional forms of the pupils in the OSH setup – this is known as *pupil-engineering* or *PSF-engineering* [34, 35]. Using pupil-engineering, coding [34], cryptography [35], recognition [36], and filtering [37] can be performed on the fly during hologram acquisition. Moreover, super resolution [38, 39] and sectional reconstruction [40, 41] can be realized by employing OSH. OSH also finds numerous applications in three-dimensional display and storage.

In this section we will only introduce the fundamental concept of OSH. The advanced properties and applications of OSH can be found in the above mentioned references or in Refs. [42–45].

5.4.1 Fundamental principles

Figure 5.14 depicts the typical setup of an optical scanning holographic system. The laser beam operated at angular frequency ω_0 is separated into two beams by a beamsplitter (BS1). The transmitted light from BS1 (beam 1) is first collimated and then passes through pupil p_1 at the front focal plane of lens L1. The frequency of the reflected light from BS1 (beam 2) is shifted to $\omega_0 + \Omega$ by an acousto-optic frequency shifter (AOFS). The light is also collimated and passes through pupil p_2 at the front focal plane of lens L1. Beam 1 and beam 2 are combined by beamsplitter BS2 in front of lens L1. Consequently, the complex field on the object plane is given by

$$S(x, y; z) = P_{1z}\left(\frac{k_0 x}{f}, \frac{k_0 y}{f}\right)\exp(j\omega_0 t) + P_{2z}\left(\frac{k_0 x}{f}, \frac{k_0 y}{f}\right)\exp[j(\omega_0 + \Omega)t], \quad (5.42)$$

where $P_{iz}(k_0 x/f,\ k_0 y/f)$ is the optical field of beam 1 or 2 at a distance z away from the back focal plane of L1, and through Fresnel diffraction [Eq. (1.35)], it is given by

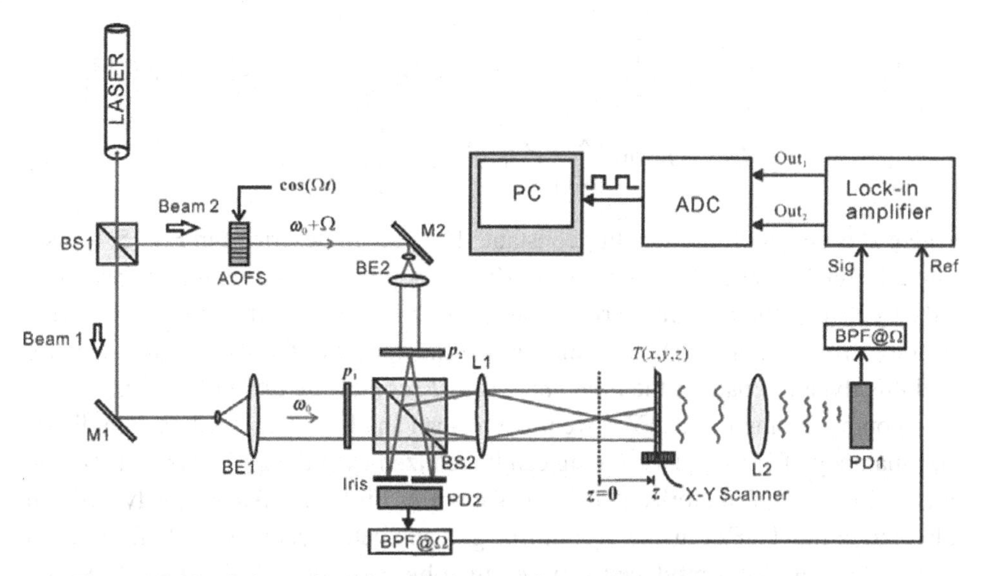

Figure 5.14 Schematic of optical scanning holography. BS beamsplitter; M mirror; BE beam expander; L lens; PD photodetector; BPF band-pass filter.

$$P_{iz}\left(\frac{k_0x}{f}, \frac{k_0y}{f}\right) = P_i\left(\frac{k_0x}{f}, \frac{k_0y}{f}\right) * h(x, y; z), \quad i = 1, 2. \tag{5.43}$$

In the above equation $P_i(k_0x/f,\ k_0y/f)$ is the optical field at the back focal plane of lens L1. According to the setup described in Fig. 5.14, and knowing that the Fourier transform of the pupil in the front focal plane of a lens is the field distribution at the back focal plane of the lens [see Eq. (1.45)], we have

$$P_i\left(\frac{k_0x}{f}, \frac{k_0y}{f}\right) = \mathcal{F}\{p_i(x, y)\}\big|_{k_x=\frac{k_0x}{f}, k_y=\frac{k_0y}{f}}, \quad i = 1, 2 \tag{5.44}$$

where $p_i(x, y)$ is the pupil function at the front focal plane of L1.

To find the complex field at the object plane, we consider a typical case, namely the first pupil is considered clear, i.e., $p_1(x, y) = 1$, and the second pupil is a pinhole, $p_2(x, y) = \delta(x, y)$. Therefore, beam 1 becomes a spherical wave behind lens L1, while beam 2 becomes a plane wave, as shown in Fig. 5.14. The optical field at z away from the focal plane of lens L1 is thus given by

$$S(x, y; z) = \frac{jk_0}{2\pi z} \exp\left[\frac{-jk_0}{2z}(x^2 + y^2)\right] \exp(j\omega_0 t) + a \exp[j(\omega_0 + \Omega)t], \tag{5.45}$$

where a is a proportionality constant and a common factor, $\exp(-jk_0z)$, has been dropped for simplicity. The above equation is similar to Eq. (2.3) for a point-source hologram. The intensity on the object plane is

$$|S(x,y;z)|^2 = \left| \frac{jk_0}{2\pi z} \exp\left[\frac{-jk_0}{2z}(x^2+y^2)\right] \exp(j\omega_0 t) + a \exp[j(\omega_0+\Omega)t] \right|^2$$
$$= DC + B \sin\left[\frac{k_0}{2z}(x^2+y^2) + \Omega t\right], \quad (5.46)$$

where B is also a proportionality constant. The pattern described in Eq. (5.46) is a static Fresnel zone plate given by Eq. (2.4) when there is no temporal frequency offset between the two interfering beams, i.e., $\Omega = 0$, but when $\Omega \neq 0$ we have running fringes that will be moving toward the center of the zone pattern and we call this dynamic pattern the *time-dependent Fresnel zone plate* (TDFZP).

Upon producing the TDFZP, we use it to scan an object target with an amplitude transparency of $T(x, y; z)$. Scanning can be realized by moving the object target on a two-dimensional motorized stage, as shown in Fig. 5.14. Alternatively, we can also move the TDFZP using x–y mirror galvanometer scanners. In both cases, a relative movement is produced between the object target and the TDFZP. At any moment of the movement, there is a shift between the object target and the TDFZP. Assume that the center of $S(x, y; z)$ is moved from $(x, y) = (0, 0)$ to $(x, y) = (x', y')$, the light transmitted through (if the object target is transparent or semi-transparent) or scattered away from (if the object target is diffusely reflecting) the object target will be proportional to $T(x, y; z)S(x - x', y - y'; z)$. The lens L2 is used to collect the transmitted or scattered light, while photodetector PD1 detects the light. The photodetector detects the intensity and integrates the intensity over its active surface, giving the output current i as

$$i(x',y') \propto \iiint |S(x,y;z)T(x+x',y+y';z)|^2 dxdydz. \quad (5.47)$$

In the above equation, we have modeled the three-dimensional object target as a collection of infinitely thin sections z away from the back focal plane of lens L1. Hence the integration over z, i.e., along the depth of the object, corresponds to the recording of a three-dimensional object. For a general case, by substituting Eq. (5.42) into Eq. (5.47), we can separate Eq. (5.47) into two terms, the baseband current (DC current) i_{DC} and the heterodyne current i_Ω:

$$i_{DC}(x',y') \propto \int |P_{1z}|^2 \otimes |T|^2 dz + \int |P_{2z}|^2 \otimes |T|^2 dz, \quad (5.48a)$$

$$i_\Omega(x',y') \propto \int P_{1z}P_{2z}^* e^{-j\Omega t} \otimes |T|^2 dz + \int P_{2z}P_{1z}^* e^{j\Omega t} \otimes |T|^2 dz, \quad (5.48b)$$

where $\otimes$ denotes correlation involving the x' and y' coordinates. We can use a bandpass filter (BPF) to filter out the DC current i_{DC}. The remaining heterodyne current i_Ω oscillating at frequency Ω is sent to a standard lock-in amplifier.

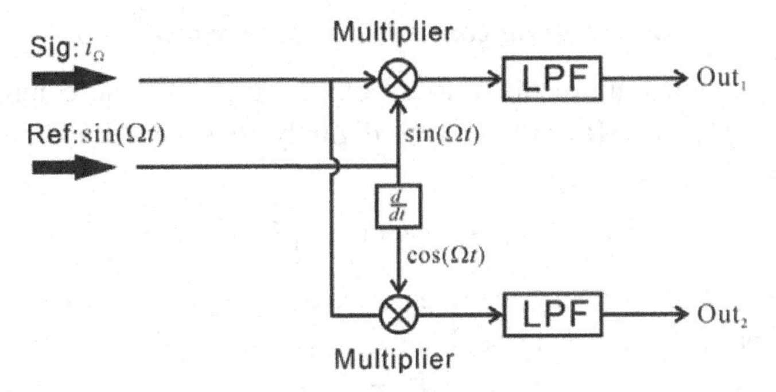

Figure 5.15 Circuit diagram of a lock-in amplifier.

A diagram of the core circuit of the lock-in amplifier is shown in Fig. 5.15. This circuit performs the so-called *synchronous demodulation.* Basically, a reference signal, sin(Ω*t*), also oscillating at frequency Ω, is applied to the lock-in amplifier to demodulate the signal of interest. In practice, the reference signal comes from another fixed photodetector PD2, as shown in Fig. 5.14. In this way, any unwanted phase fluctuation within the interferometer (formed by beamsplitters BS1, BS2, and mirrors M1 and M2) can be compensated significantly.

For a general dual-channel lock-in amplifier, there are two outputs given by

$$\mathrm{Out}_1 \propto \mathrm{LPF}\{i_\Omega \times \sin(\Omega t)\}, \tag{5.49a}$$

and

$$\mathrm{Out}_2 \propto \mathrm{LPF}\{i_\Omega \times \cos(\Omega t)\}, \tag{5.49b}$$

where LPF{•} stands for the operation of lowpass filtering, i.e., any frequency of Ω or higher will be filtered out and will not be present at the output of the filter. For convenience, we let $P = P_{1z}P_{2z}^* = |P|\exp(j\Phi)$. Thus Eq. (5.48b) can be re-written as

$$i_\Omega = \mathrm{Re}\left\{\int P \otimes |T|^2 e^{-j\Omega t} dz\right\} = \int |P| \cos(\Phi - \Omega t) \otimes |T|^2 dz. \tag{5.50}$$

By substituting Eq. (5.50) into Eqs. (5.49a) and (5.49b) and evoking the operation of lowpass filtering, we have

$$\mathrm{Out}_1 = \int |P| \sin\Phi \otimes |T|^2 dz, \tag{5.51a}$$

$$\mathrm{Out}_2 = \int |P| \cos\Phi \otimes |T|^2 dz, \tag{5.51b}$$

for the outputs of the lock-in amplifier. The two outputs can be converted to digital signals by an analog-to-digital converter (ADC), and therefore can be transmitted or processed digitally.

5.4.2 Hologram construction and reconstruction

To construct a hologram of the scanned object target, we can combine the two outputs from Eqs. (5.51a) and (5.51b) digitally to generate a digital complex hologram H_c as

$$H_c = \text{Out}_2 + j \times \text{Out}_1 = \int P \otimes |T|^2 dz. \tag{5.52}$$

To see the effect of retrieving the three-dimensional location of the object target, we assume that the object is located at $z = z_0$ and its amplitude transmittance can be expressed as $T(x, y; z) = T(x, y) \times \delta(z - z_0)$. Hence Eq. (5.52) can be reduced to

$$H_c(x,y) = P(x,y;z_0) \otimes |T(x,y)|^2, \tag{5.53}$$

where $P(x,y;z_0) = P_{1z_0}P^*_{2z_0}$. In the typical case described in Fig. 5.14, beam 1 is a spherical wave on the object target, i.e., $p_1(x, y) = 1$, and beam 2 is a plane wave, i.e., $p_2(x, y) = \delta(x, y)$, and according to Eq. (5.43) we have

$$P(x,y;z_0) = \hat{h}(x,y;z_0) = \frac{k_0}{2\pi z}\exp\left[\frac{-jk_0}{2z}(x^2+y^2)\right]. \tag{5.54}$$

As a result, Eq. (5.53) can be expressed as a complex Fresnel zone plate hologram given by

$$H_c(x,y) = \hat{h}(x,y;z_0) \otimes |T(x,y)|^2. \tag{5.55}$$

We can easily identify $H_c(x, y)$ as the *complex hologram* of $|T(x, y)|^2$. Thus we can apply the Fresnel diffraction formula [Eq. (1.35)] to retrieve the field distribution at any plane of interest away from the hologram, i.e.,

$$\psi_p(x,y;z_r) \propto \hat{h}(x,y;z_r) * \left[\hat{h}(x,y;z_0) \otimes |T(x,y)|^2\right], \tag{5.56}$$

where z_r is the distance measured from the hologram plane. Apparently, the intensity distribution of the object target can be reconstructed at $z_r = z_0$ because $\psi_p(x, y; z_0) = |T(x, y)|^2$. This is a real image of the reconstruction [see Problem 5.5]. If we reconstruct $H_c^*(x,y)$, the reconstruction is a virtual image.

It is noted that the reconstruction process of optical scanning holography (OSH) is exactly the same as that in conventional holography. But in OSH there is no annoying zeroth-order beam and twin image among the reconstructed field. This is the main merit of the complex hologram. Also we note that in the above process only the intensity distribution of the object target $|T(x, y)|^2$ is retrieved. Indeed, because the phase of the object $T(x, y)$ is not recorded, OSH is operated in the *incoherent mode*. However, the phase of the object target can also be recorded, provided a pinhole mask is attached to photodetector PD1 located at the back focal plane of lens L2 [46]. We will not discuss the details here but only show the results. The resulting two outputs become

$$\text{Out}_1 = \text{Im}\left\{\int P \otimes T\, dz\right\}, \tag{5.57a}$$

$$\text{Out}_2 = \text{Re}\left\{\int P \otimes T\, dz\right\}. \tag{5.57b}$$

Similar to the incoherent mode, we can combine the two outputs to produce H_c,

$$H_c = \text{Out}_2 + j \times \text{Out}_1 = \int P \otimes T \ dz. \tag{5.58}$$

It is noted that now H_c is the complex hologram of T and OSH is operated in the *coherent mode*. Figure 5.16 shows the reconstruction of a complex hologram given by Eq. (5.58), which is recorded in the coherent mode [47], while in Fig. 5.17 we show the holograms and reconstruction of a diffusely reflecting object (a die), illustrating the incoherent operation of optical scanning holography as the reconstruction is free of the speckle noise commonly exhibited in coherent optical systems [48]. Figure 5.17(a) and (b) are the real and imaginary parts of the complex hologram given by Eq. (5.52), respectively, where the real part of the hologram is given by Eq. (5.51b) and the imaginary part is given by Eq. (5.51a). Digital reconstruction of the complex hologram is shown in Fig. 5.17(c), which is obviously not contaminated by speckle noise as compared to the image shown in Fig. 5.17(d) where the image is observed by a CCD camera upon coherent illumination of the die. Reference [49] is the most recent review paper on optical scanning holography.

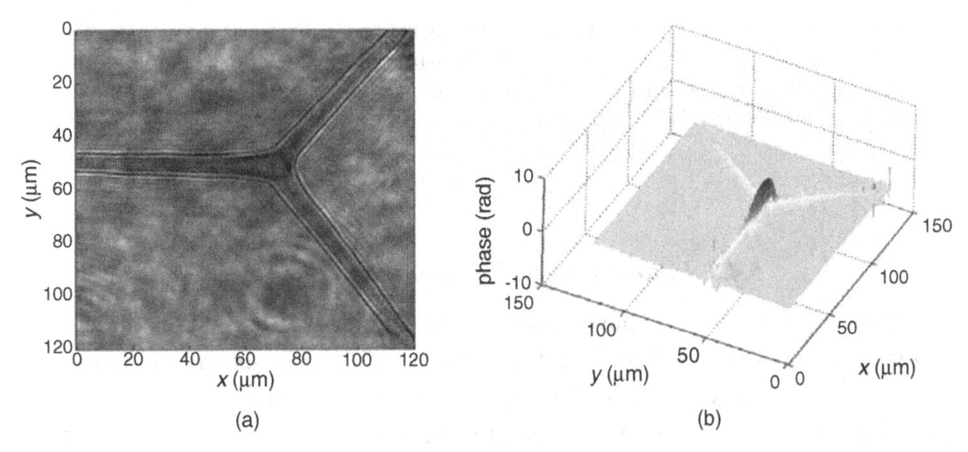

Figure 5.16 Reconstruction of a hologram recorded in coherent mode. The specimen is a good phase object: siliceous three-pronged spongilla spicule. (a) Absolute value of the reconstruction amplitude, and (b) three-dimensional phase profile of the specimen. Reprinted from Ref. [47], with permission.

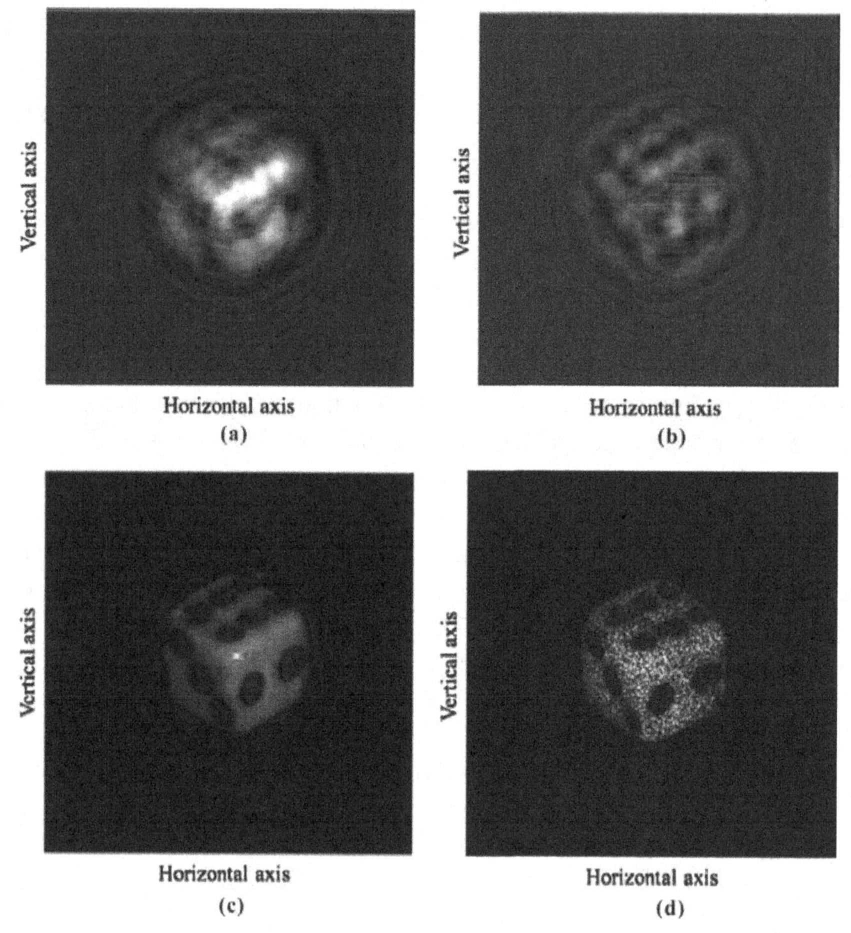

Figure 5.17 (a) Real part of a complex hologram of a die, (b) imaginary part of a complex hologram of a die, (c) digital reconstruction of the complex hologram, and (d) image observed by a CCD camera upon coherent illumination of the die. Reprinted from Ref. [48], with permission, © OSA.

5.4.3 Intuition on optical scanning holography

In the above subsection we have described the setup for OSH and the corresponding mathematical model. However, the operation of OSH is not straightforward to those who initially study OSH. In this subsection we would like to explain the principle of OSH from an intuitive point of view.

We first return to conventional holography. In the recording process as shown in Fig. 5.18(a), we use coherent light to illuminate an object. Ideally, each object point produces spherical wave scattered light toward the hologram plane, a plane

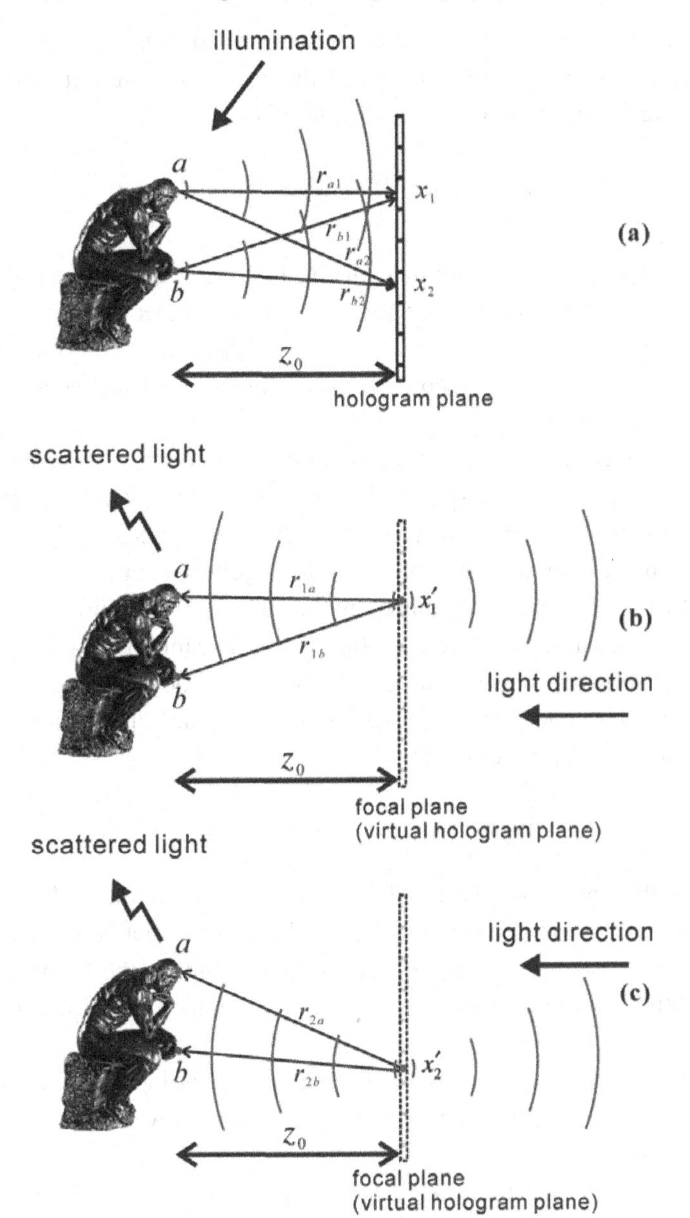

Figure 5.18 Recording for (a) a conventional hologram, (b) a virtual hologram scanned at x_1', and (c) a virtual hologram scanned at x_2'.

where the interference patterns are recorded. Let us consider a single pixel, e.g. x_1, on the hologram plane. The object light on the pixel, ψ_1, is the superposition of the light emerging from all the object points, that is

$$\psi_1 = \psi_{a0} e^{-jkr_{a1}} + \psi_{b0} e^{-jkr_{b1}} + \cdots, \tag{5.59}$$

where ψ_{a0} is the complex amplitude of the light emerging from object point a, which is dependent on the surface properties of the object; r_{a1} is the path from object point a to hologram pixel x_1, and so on. So the three-dimensional information of the object can be recorded via the interference between the scattered light and a reference light.

Now we consider a comparable setup in OSH. Although in typical OSH we use the interference pattern of a spherical wave and a plane wave to illuminate the object, we ignore the plane wave in the discussion because it serves as the reference light. So how do we apply OSH to acquire a data point on a hologram, the same as the data point obtained in the hologram at point x_1 as shown in Fig. 5.18(a)? We can illuminate the object with a scanning beam from the right side, and let a spherical wave focus at location x_1', as shown in Fig. 5.18(b). Then, the light on the photodetector, ψ_{S_1}, is proportional to the superposition of the light emerging from all object points, that is

$$\psi_{S_1} \propto \psi_{a0} e^{-jkr_{1a}} + \psi_{b0} e^{-jkr_{1b}} + \cdots, \tag{5.60}$$

where r_{1a} is the path from the focal point x_1' to object point a. Assume that the geometrical relationship between a and x_1' is the same as that between a and x_1, we can say $r_{1a} = r_{a1}$ and $\psi_{S_1} \propto \psi_1$ so that the data point on the hologram obtained in Fig. 5.18(b) is proportional to the data point obtained at hologram point x_1 in Fig. 5.18(a).

Similarly, we can obtain another data point on the hologram comparable to the data point at x_2 in Fig. 5.18(a) by moving the spherical wave to focus at location x_2', as shown in Fig. 5.18(c).

In conclusion, we find that the back focal plane of lens L1 in Fig. 5.14 can be regarded as a virtual hologram plane in comparison with conventional holography. As we scan the object using the interference pattern between a spherical wave and a plane wave, we acquire the holographic data pixel by pixel. Since the hologram data obtained in Fig. 5.18(a) and those obtained in Fig. 5.18(b) and (c) are the same, the reconstruction method is also the same: we only need to back-propagate the object light from the hologram plane to the object plane, which is z_0 in both cases.

Problems

5.1 Show that, in three-step phase-shifting holography, the complex amplitude of the object is given by

$$\psi_0 = \frac{(1+j)\left(I_0 - I_{\pi/2}\right) + (j-1)\left(I_\pi - I_{\pi/2}\right)}{4\psi_r^*},$$

where the three holograms are

$$I_\delta = |\psi_0 + \psi_r \exp(-j\delta)|^2$$

with phase steps $\delta = 0$, $\delta = \pi/2$, and $\delta = \pi$. ψ_r is the complex amplitude of the reference light.

5.2 If the phase steps in three-step phase-shifting holography are 0, δ_1, and δ_2, find the complex amplitude of the object, ψ_0, in terms of two intensity maps, i.e., $|\psi_0|^2$ and $|\psi_r|^2$, three holograms, and two phase steps δ_1 and δ_2. ψ_r is the complex amplitude of the reference light.

5.3 Prove that for two-step phase-shifting holography [Eq. (5.16)], $\pi/2$ is the phase step that is most insensitive to the phase error.

5.4 The heterodyne current from optical scanning holography is given by Eq. (5.50) as follows:

$$i_\Omega = \mathrm{Re}\left\{\int P \otimes |T|^2 e^{-j\Omega t} dz\right\}.$$

Show that the two outputs from a dual channel lock-in amplifier are given by

$$\mathrm{Out}_1 = \int |P| \sin\Phi \otimes |T|^2 dz,$$

$$\mathrm{Out}_2 = \int |P| \cos\Phi \otimes |T|^2 dz.$$

5.5 (a) Prove that

$$f_1(x) \otimes f_2(x) = f_1^*(-x) * f_2(x),$$

where $\otimes$ and $*$ are the operators of correlation and convolution, respectively, and defined according to Table 1.1.

(b) Apply the result in part (a), and prove that

$$\hat{h}(x, y; z_0) * \left[\hat{h}(x, y; z_0) \otimes |T(x, y)|^2\right] = |T(x, y)|^2.$$

References

1. C. B. Burckhardt, and L. H. Enloe, Television transmission of holograms with reduced resolution requirements on the camera tabes, *Bell Syst. Tech. J.* **48**, 1529–1535 (1969).
2. T. Zhang, and I. Yamaguchi, Three-dimensional microscopy with phase-shifting digital holography, *Optics Letters* **23**, 1221–1223 (1998).
3. I. Yamaguchi, J.-I. Kato, S. Ohta, and J. Mizuno, Image formation in phase-shifting digital holography and applications to microscopy, *Applied Optics* **40**, 6177–6186 (2001).
4. J. Rosen, G. Indebetouw, and G. Brooker, Homodyne scanning holography, *Optics Express* **14**, 4280–4285 (2006).
5. M. Gross, and M. Atlan, Digital holography with ultimate sensitivity, *Optics Letters* **32**, 909–911 (2007).
6. P. Guo, and A. J. Devaney, Digital microscopy using phase-shifting digital holography with two reference waves, *Optics Letters* **29**, 857–859 (2004).
7. J.-P. Liu, T.-C. Poon, G.-S. Jhou, and P.-J. Chen, Comparison of two-, three-, and four-exposure quadrature phase-shifting holography, *Applied Optics* **50**, 2443–2450 (2011).
8. M. Atlan, M. Gross, and E. Absil, Accurate phase-shifting digital interferometry, *Optics Letters* **32**, 1456–1458 (2007).
9. V. Micó, J. García, Z. Zalevsky, and B. Javidi, Phase-shifting Gabor holography, *Optics Letters* **34**, 1492–1494 (2009).
10. J.-P. Liu, and T.-C. Poon, Two-step-only quadrature phase-shifting digital holography, *Optics Letters* **34**, 250–252 (2009).
11. X. F. Meng, L. Z. Cai, X. F. Xu, X. L. Yang, X. X. Shen, G. Y. Dong, and Y. R. Wang, Two-step phase-shifting interferometry and its application in image encryption, *Optics Letters* **31**, 1414–1416 (2006).
12. X. F. Meng, X. Peng, L. Z. Cai, A. M. Li, J. P. Guo, and Y. R. Wang, Wavefront reconstruction and three-dimensional shape measurement by two-step dc-term-suppressed phase-shifted intensities, *Optics Letters* **34**, 1210–1212 (2009).
13. C.-S. Guo, L. Zhang, H.-T. Wang, J. Liao, and Y. Y. Zhu, Phase-shifting error and its elimination in phase-shifting digital holography, *Optics Letters* **27**, 1687–1689 (2002).
14. W. Chen, C. Quan, C. J. Tay, and Y. Fu, Quantitative detection and compensation of phase-shifting error in two-step phase-shifting digital holography, *Optics Communications* **282**, 2800–2805 (2009).
15. Y. Awatsuji, T. Tahara, A. Kaneko, T. Koyama, K. Nishio, S. Ura, T. Kubota, and O. Matoba, Parallel two-step phase-shifting digital holography, *Applied Optics* **47**, D183–D189 (2008).
16. Y. Awatsuji, A. Fujii, T. Kubota, and O. Matoba, Parallel three-step phase-shifting digital holography, *Applied Optics* **45**, 2995–3002 (2006).
17. T. Tahara, K. Ito, T. Kakue, M. Fujii, Y. Shimozato, Y. Awatsuji, K. Nishio, S. Ura, T. Kubota, and O. Matoba, Parallel phase-shifting digital holographic microscopy, *Biomedical Optics Express* **1**, 610–616 (2010).
18. T. Kakue, Y. Moritani, K. Ito, Y. Shimozato, Y. Awatsuji, K. Nishio, S. Ura, T. Kubota, and O. Matoba, Image quality improvement of parallel four-step phase-shifting digital holography by using the algorithm of parallel two-step phase-shifting digital holography, *Optics Express* **18**, 9555–9560 (2010).
19. T. Tahara, K. Ito, M. Fujii, T. Kakue, Y. Shimozato, Y. Awatsuji, K. Nishio, S. Ura, T. Kubota, and O. Matoba, Experimental demonstration of parallel two-step phase-shifting digital holography, *Optics Express* **18**, 18975–18980 (2010).

20. T. Nomura, and M. Imbe, Single-exposure phase-shifting digital holography using a random-phase reference wave, *Optics Letters* **35**, 2281–2283 (2010).
21. Y.-C. Lin, C.-J. Cheng, and T.-C. Poon, Optical sectioning with a low-coherence phase-shifting digital holographic microscope, *Applied Optics* **50**, B25–B30 (2011).
22. P. Massatsch, F. Charrière, E. Cuche, P. Marquet, and C. D. Depeursinge, Time-domain optical coherence tomography with digital holographic microscopy, *Applied Optics* **44**, 1806–1812 (2005).
23. S. Tamano, Y. Hayasaki, and N. Nishida, Phase-shifting digital holography with a low-coherence light source for reconstruction of a digital relief object hidden behind a light-scattering medium, *Applied Optics* **45**, 953–959 (2006).
24. G. Pedrini, and H. J. Tiziani, Short-coherence digital microscopy by use of a lensless holographic imaging system, *Applied Optics* **41**, 4489–4496 (2002).
25. E. Wolf, Three-dimensional structure determination of semi-transparent objects from holographic data, *Optics Communications* **1**, 153–156 (1969).
26. Y. Sung, W. Choi, C. Fang-Yen, K. Badizadegan, R. R. Dasari, and M. S. Feld, Optical diffraction tomography for high resolution live cell imaging, *Optics Express* **17**, 266–277 (2009).
27. J. J. Stamnes, Focusing of a perfect wave and the airy pattern formula, *Optics Communications* **37**, 311–314 (1981).
28. T.-C. Poon, and A. Korpel, Optical transfer function of an acousto-optic heterodyning image processor, *Optics Letters* **4**, 317–319 (1979).
29. T.-C. Poon, Scanning holography and two-dimensional image processing by acousto-optic two-pupil synthesis, *Journal of the Optical Society of America A* **2**, 521–527 (1985).
30. G. Indebetouw, T. Kim, T.-C. Poon, and B. W. Schilling, Three-dimensional location of fluorescent inhomogeneities in turbid media by scanning heterodyne holography, *Optics Letters* **23**, 135–137 (1998).
31. B. W. Schilling, T.-C. Poon, G. Indebetouw, B. Storrie, K. Shinoda, Y. Suzuki, and M. H. Wu, Three-dimensional holographic fluorescence microscopy, *Optics Letters* **22**, 1506–1508 (1997).
32. G. Indebetouw, and W. Zhong, Scanning holographic microscopy of three-dimensional fluorescent specimens, *Journal of the Optical Society of America A* **23**, 1699–1707 (2006).
33. G. Indebetouw, A posteriori quasi-sectioning of the three-dimensional reconstructions of scanning holographic microscopy, *Journal of the Optical Society of America A* **23**, 2657–2661 (2006).
34. Y. Shinoda, J.-P. Liu, P. Sheun Chung, K. Dobson, X. Zhou, and T.-C. Poon, Three-dimensional complex image coding using a circular Dammann grating, *Applied Optics* **50**, B38–B45 (2011).
35. T.-C. Poon, T. Kim, and K. Doh, Optical scanning cryptography for secure wireless transmission, *Applied Optics* **42**, 6496–6503 (2003).
36. T.-C. Poon, and T. Kim, Optical image recognition of three-dimensional objects, *Applied Optics* **38**, 370–381 (1999).
37. J.-P. Liu, C.-C. Lee, Y.-H. Lo, and D.-Z. Luo, Vertical-bandwidth-limited digital holography, *Optics Letters* **37**, 2574–2576 (2012).
38. G. Indebetouw, A. El Maghnouji, and R. Foster, Scanning holographic microscopy with transverse resolution exceeding the Rayleigh limit and extended depth of focus, *Journal of the Optical Society of America A* **22**, 892–898 (2005).

39. G. Indebetouw, Y. Tada, J. Rosen, and G. Brooker, Scanning holographic microscopy with resolution exceeding the Rayleigh limit of the objective by superposition of off-axis holograms, *Applied Optics* **46**, 993–1000 (2007).
40. H. Kim, S.-W. Min, B. Lee, and T.-C. Poon, Optical sectioning for optical scanning holography using phase-space filtering with Wigner distribution functions, *Applied Optics* **47**, D164–D175 (2008).
41. E. Y. Lam, X. Zhang, H. Vo, T.-C. Poon, and G. Indebetouw, Three-dimensional microscopy and sectional image reconstruction using optical scanning holography, *Applied Optics* **48**, H113–H119 (2009).
42. T.-C. Poon, *Optical Scanning Holography with MATLAB* (Springer, New York, 2007).
43. G. Indebetouw, The multi-functional aspect of scanning holographic microscopy: a review, *Chinese Optics Letters* **7**, 1066–1071 (2009).
44. T.-C. Poon, Recent progress in optical scanning holography, *Journal of Holography and Speckle* **1**, 6–25 (2004).
45. T.-C. Poon, Optical scanning holography: principles and applications, in *Three-Dimensional Holographic Imaging*, C. J. Kuo, and M. H. Tsai, eds. (John Wiley & Sons, New York, 2002), pp. 49–75.
46. T.-C. Poon, and G. Indebetouw, Three-dimensional point spread functions of an optical heterodyne scanning image processor, *Applied Optics* **42**, 1485–1492 (2003).
47. G. Indebetouw, Y. Tada, and J. Leacock, Quantitative phase imaging with scanning holographic microscopy: an experimental assessment, *BioMedical Engineering OnLine* **5**, 63 (2006).
48. Y. S. Kim, T. Kim, S. S. Woo, H. Kang, T.-C. Poon, and C. Zhou, Speckle-free digital holographic recording of a diffusely reflecting object, *Optics Express* **21**, 8183–8189 (2013).
49. T.-C. Poon, Optical scanning holography – a review of recent progress, *Journal of the Optical Society of Korea* **13**, 406–415 (2009).

6
Applications in digital holography

In this chapter, we discuss some of the important and modern applications in digital holography, including holographic microscopy, sectioning, phase extraction, optical contouring, and deformation measurements.

6.1 Holographic microscopy

The first Gabor hologram was invented as a new microscopic technique [1], but it suffers from the problem of twin image. In 1965, a successful holographic microscope was demonstrated by E. Leith and J. Upatnieks [2]. However, the use of holographic microscopy is limited by its inconvenient chemical developing procedures. This drawback is apparently disappearing in the era of digital holography. Digital holographic microscopy (DHM) has unique merits in comparison with traditional optical microscopy. DHM can acquire the three-dimensional information of a sample simultaneously, and the phase of the sample can be determined quantitatively. DHM has become one of the most important applications in digital holography. In DHM, some effort must be made to achieve micrometer or submicrometer resolution. According to the recording geometry, DHM can be categorized into three types [OSH, see Section 5.4, is not taken into account here]: microscope-based DHM [3–8], Fourier-based DHM [9–14], and spherical-reference-based DHM [15]. In the following sections, we discuss the three kinds of DHM.

6.1.1 Microscope-based digital holographic microscopy

Figure 6.1 depicts the setup of a typical transmission-type microscope-based digital holographic microscope. The mirror (M) in the reference arm can be tilted for off-axis holographic recording, or can be movable for phase-shifting recordings. In the specimen arm, a microscope objective (MO) is used to produce a magnified image of the specimen. The digital holographic recording geometry can

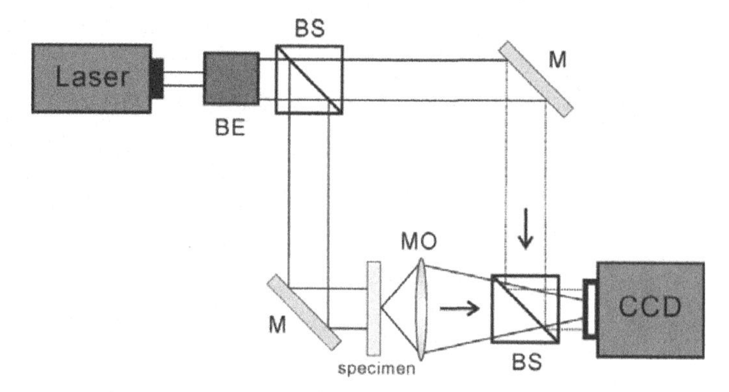

Figure 6.1 Setup of a transmission-type microscope-based digital holographic microscope. BE beam expander; BS beamsplitter; M mirror; MO microscope objective.

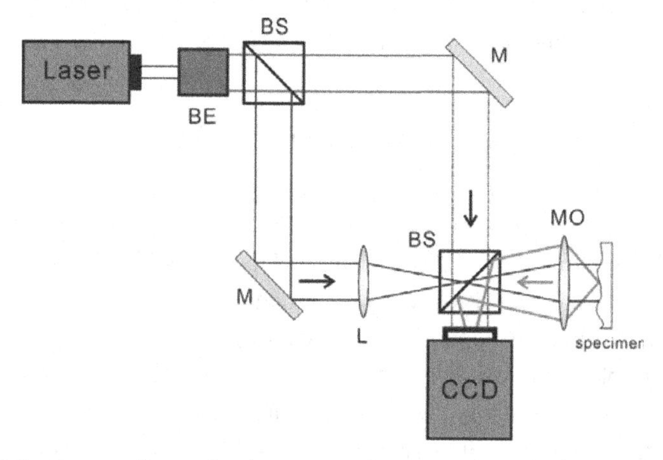

Figure 6.2 Setup of a reflection-type microscope-based digital holographic microscope. BE beam expander; BS beamsplitter; M mirror; L lens; MO microscope objective.

also be realized as a reflection-type setup, as shown in Fig. 6.2. In the reflection-type setup, the object light is back-scattered to provide the object field to the CCD. Because the MO is in the path of the back-scattered object field, lens L is used additionally to provide uniform illumination of the specimen.

For both cases, the geometrical relation between the specimen and its image is shown in Fig. 6.3(a). Note that we do not need to focus the image on the CCD plane because the image is holographically recorded. Under the paraxial approximation, the lateral magnification of the image is defined as

$$\mathrm{M}_{Lat} = -\frac{d_i}{d_o} = -\frac{\theta_o}{\theta_i}, \tag{6.1}$$

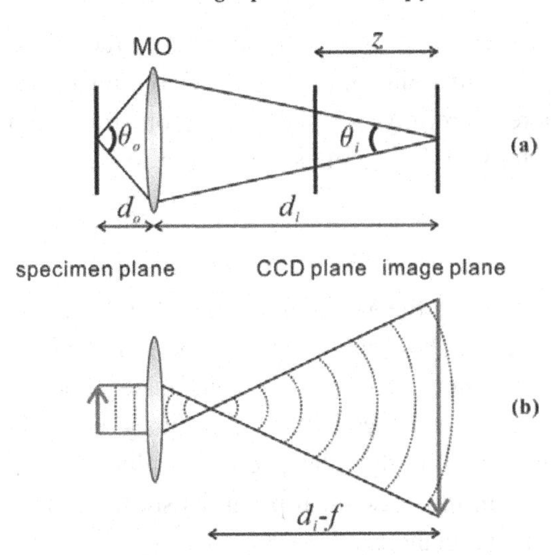

Figure 6.3 (a) Ray configuration, and (b) wavefront imaging using the microscope objective.

where d_o and d_i are the distances between the specimen and the MO, and between the MO and the image, respectively; θ_o and θ_i are the divergence angles of the object light and the image light, respectively. The angle θ_i will be much less than θ_o, provided $|\mathrm{M}_{Lat}|$ is large enough owing to a high numerical aperture (NA) of the MO. Thus the light wave scattered from the image space can be holographically recorded more easily than the light wave scattered from the object space. The main purpose of the high-NA microscope objective is to make θ_i smaller such that diffraction of the magnified image interfering with the reference wave will make coarser fringes, thereby adapting to the low resolution of the CCD. Another way to draw the same conclusion is that since the sampling period of the optical field is the pixel pitch of the CCD, $\Delta_{x_{CCD}}$, a magnified image is equivalent to having the sampling period reduced to $\Delta_{x_{CCD}}/|\mathrm{M}_{Lat}|$ in the object space. Hence the optical resolution of the MO dominates the practical resolution of the reconstructed image.

As the digital hologram is recorded, the complex field $\psi_{i0}(x, y)$ at the CCD plane can be retrieved with standard filtering or phase-shifting procedures. Subsequently, the complex field at the image plane $\psi_{iz}(x, y)$ can be obtained directly by Fresnel diffraction [Eq. (1.37)],

$$\psi_{iz}(x,y) = \exp(-jk_0 z)\frac{jk_0}{2\pi z}\exp\left[\frac{-jk_0}{2z}(x^2+y^2)\right]$$
$$\times\mathcal{F}\left\{\psi_{i0}(x,y)\exp\left[\frac{-jk_0}{2z}(x^2+y^2)\right]\right\}\Bigg|_{k_x=\frac{k_0 x}{z},\,k_y=\frac{k_0 y}{z}}. \tag{6.2}$$

However, since conventional digital holographic recordings like that being discussed are coherent imaging, the phase distribution of the magnified image needs to be considered carefully. According to geometrical optics and referring to Fig. 6.3(a), we can find the image plane according to

$$\frac{1}{d_o}+\frac{1}{d_i}=\frac{1}{f}, \tag{6.3}$$

where f is the focal length of the MO. On the other hand, wavefront imaging by the MO is shown in Fig. 6.3(b), where we assume a flat object wavefront being coherently imaged. We note that on the image plane, there is a quadratic phase error. This phase error can be calculated rigorously using Fourier optics [16]. Intuitively, we can say that the phase error is due to a divergent spherical wavefront emitting from the back focal point, as shown in Fig. 6.3(b). Hence the phase error on the image plane can be expressed as

$$\exp\left[\frac{-jk_0}{2(d_i-f)}\left(x^2+y^2\right)\right]. \tag{6.4}$$

As a result, we can multiply the field $\psi_{iz}(x, y)$ by a phase mask to compensate for the inherent phase error. The corrected complex field of the image is thus given by [4, 5]:

$$\psi_{iz}^{c}(x,y)=\psi_{iz}(x,y)\times PM, \tag{6.5}$$

where

$$PM=\exp\left[\frac{jk_0}{2L}\left(x^2+y^2\right)\right], \tag{6.6}$$

and $L=d_i-f=d_i^2/(d_i+d_o)$. With the help of Fig. 6.3(b), we know that the inherent phase error can be compensated not only at the image plane but also at the CCD plane. In addition, we can apply other phase masks, obtained digitally or optically, to compensate for any high-order aberrations in the reconstruction procedure [6, 17, 18].

6.1.2 Fourier-based digital holographic microscopy

Figure 6.4 depicts the setup of a typical Fourier-based digital holographic microscope. The object and the reference light are arranged in the geometry of lensless Fourier holography (Fig. 3.9). In microscope-based DHM, the object light is manipulated to alleviate the problem caused by the low resolution of the recording CCD. In Fourier-based DHM, the reference point source and the object are located on the same plane and the distance of this plane to the CCD is manipulated. Because at the CCD plane the curvatures of the object light and the reference light

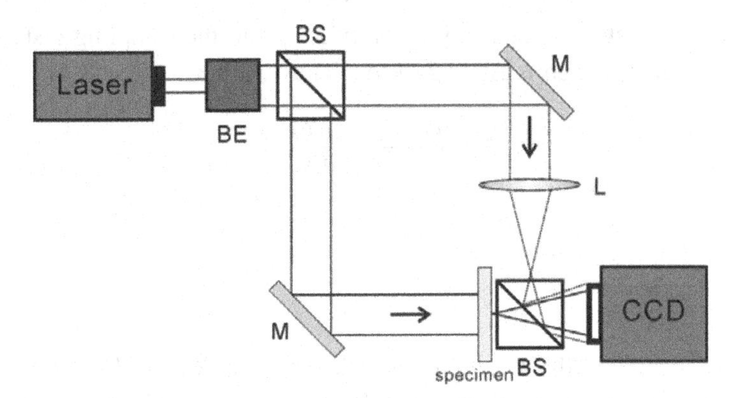

Figure 6.4 Setup of a Fourier-based digital holographic microscope. BE beam expander; BS beamsplitter; M mirror; L lens.

are matched, a higher spatial frequency of the object light can possibly be recorded. We assume that the reference light is at the optical axis, and that phase shifting is performed to remove the zeroth-order light and the twin image. The complex hologram H_c is thus expressed as

$$H_c(x,y) = \mathcal{F}\left\{\psi_{p0}(x,y)\exp\left[\frac{-jk_0}{2z_0}\left(x^2+y^2\right)\right]\right\}_{k_x=\frac{k_0x}{z_0},\,k_y=\frac{k_0y}{z_0}}. \tag{6.7}$$

We used Eq. (3.15) to obtain the above equation with $x_0 = 0$ due to the assumed on-axis point reference source and we have also eliminated the twin image through the action of phase shifting. In Eq. (6.7), $\psi_{p0}(x, y)$ is the complex field of the object, and z_0 is the distance between the object and the CCD. The complex field of the object, $\psi_{p0}(x, y)$, can be reconstructed digitally by taking the inverse transform of Eq. (6.7) to obtain

$$\psi_{p0}(x,y) = \exp\left[\frac{jk_0}{2z_0}\left(x^2+y^2\right)\right]\times\mathcal{F}^{-1}\left\{H_c\left(\frac{k_xz_0}{k_0},\frac{k_yz_0}{k_0}\right)\right\}. \tag{6.8}$$

Note that the sampling of the hologram by the CCD is in the Fourier domain of the object. Considering one dimension for simplicity, k_x and x are related by $k_x = k_0x/z_0$ according to Eq. (6.7). Hence

$$\Delta_k = \frac{k_0\Delta_{xCCD}}{z_0},$$

where Δ_{xCCD} is the pixel pitch of the CCD. On the other hand, as the spectrum is transformed to the spatial domain (reconstruction plane) by FFT, there is a relationship between the frequency resolution Δ_k and the sampling distance Δ_x in the spatial domain, which is $\Delta_k = 2\pi/M\Delta_x$. Thus we have

$$\frac{k_0 \Delta_{xCCD}}{z_0} = \frac{2\pi}{M\Delta_x}, \tag{6.9a}$$

or

$$\Delta_x = \frac{z_0 \lambda_0}{D_x}, \tag{6.9b}$$

where $M\Delta_{xCCD} = D_x$ with M being the number of samples and D_x the width of the CCD. Since Δ_x is the sampling period in the reconstruction plane, for a given CCD chip, a short object distance, z_0, is required to achieve good resolution. Now $2\pi/\Delta_k = M\Delta_x = z_0\lambda_0/\Delta_{xCCD}$ gives the size of the reconstructed field (also the illumination area of the object).

6.1.3 Spherical-reference-based digital holographic microscopy

In the above analysis, the focus spot of the reference light is assumed to be at the same plane as the specimen. If this is not the case, Eq. (6.7) cannot be applied directly, and the system becomes a spherical-reference-based digital holographic microscope. The setup is similar to that shown in Fig. 6.4, and the relationship between the specimen and the reference light is illustrated in Fig. 6.5(a). The complex hologram obtained is expressed as

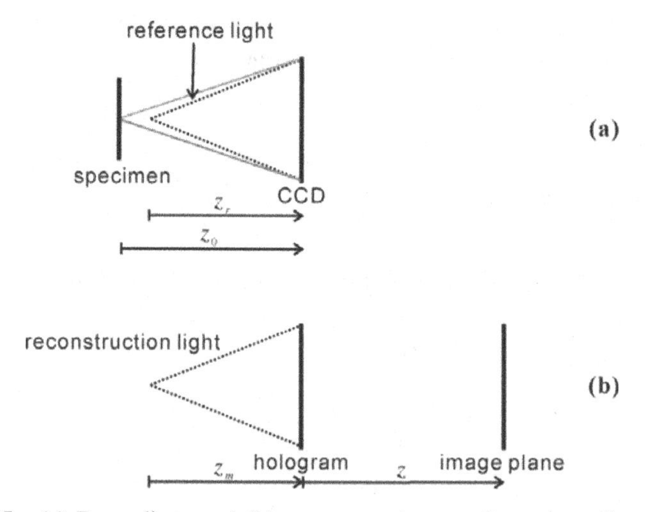

Figure 6.5 (a) Recording, and (b) reconstruction configurations for spherical-reference-light DHM.

$$H_c(x,y) = \exp\left[\frac{-jk_0}{2}\left(\frac{1}{z_0}-\frac{1}{z_r}\right)(x^2+y^2)\right]$$
$$\times \mathcal{F}\left\{\psi_{p0}(x,y)\exp\left[\frac{-jk_0}{2z_0}(x^2+y^2)\right]\right\}_{k_x=\frac{k_0 x}{z_0},\, k_y=\frac{k_0 y}{z_0}}, \tag{6.10}$$

where z_0 and z_r are the distances between the specimen and the CCD and between the reference point source and the CCD, respectively. As the complex hologram is obtained, we recommend first interpolating the hologram before multiplying by a digital spherical wave, $R_m(x, y)$, to reconstruct the complex hologram. As shown in Fig. 6.5(b), the spherical reconstruction light is given by

$$R_m(x,y) = \exp\left[\frac{-jk_0}{2z_m}(x^2+y^2)\right], \tag{6.11}$$

where z_m is the distance between the reconstruction point source and the CCD. Finally, we perform free-space propagation to obtain the reconstructed image. We can set $z_m = z_r$, resulting in a field at the hologram plane as

$$\psi(x,y) = \exp\left[\frac{-jk_0}{2z_0}(x^2+y^2)\right]$$
$$\times \mathcal{F}\left\{\psi_{p0}(x,y)\exp\left[\frac{-jk_0}{2z_0}(x^2+y^2)\right]\right\}_{k_x=\frac{k_0 x}{z_0},\, k_y=\frac{k_0 y}{z_0}}. \tag{6.12}$$

Note that Eq. (6.12) is in the form of Fresnel diffraction. Thus the object light at the specimen plane $\psi_{p0}(x, y)$ can be retrieved from backward propagation by setting $z = -z_0$ (Section 4.3.5). As described in Section 2.3, we can also set $z_m \neq z_r$ to introduce a magnification of the reconstructed image [15]. However, the resolution achieved of the specimen is still limited to $\lambda_0 z_0/D_x$ (Section 4.2), where D_x is the size of the CCD. By using a spherical reference light, the distance from the specimen to the CCD can be significantly reduced. Thus high-resolution reconstruction can be achieved.

Besides the above mentioned techniques, Gabor holography can also be applied to DHM. Figure 6.6 shows a typical setup of a Gabor digital holographic microscope. We assume that the specimen is illuminated with a spherical wave. The spherical wave itself serves as the reference wave, and the light scattered from the small particles in the specimen is the object wave. The setup is very simple, and the relationship between the object wave and the reference wave is similar to that of spherical-reference-based DHM. As a result, we can apply the same reconstruction method to the hologram obtained by a Gabor digital holographic microscope. Indeed, the zeroth-order light and the twin image associated with the Gabor hologram can be removed by phase-shifting

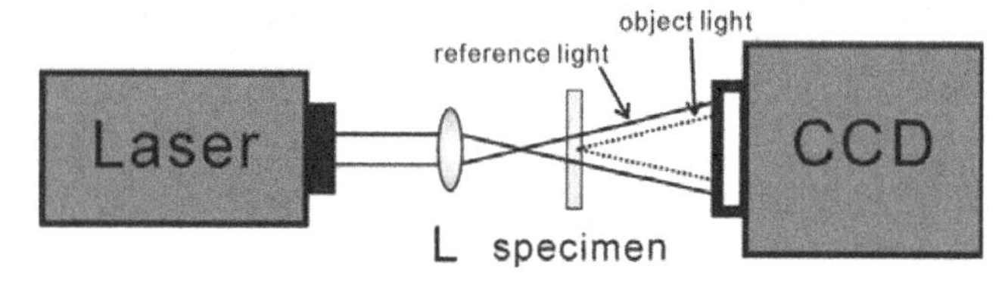

Figure 6.6 Setup for Garbor holography DHM.

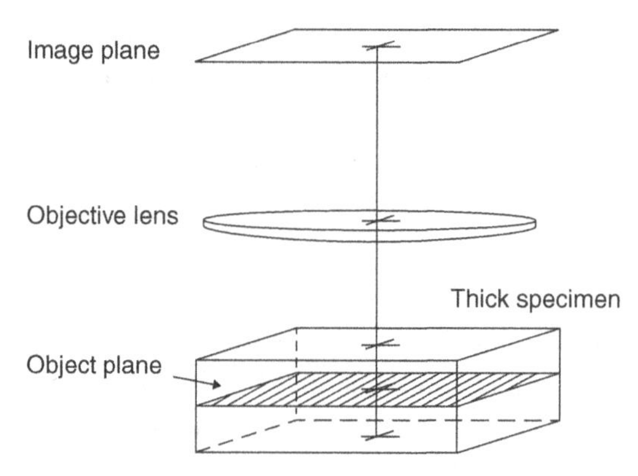

Figure 6.7 Imaging a three-dimensional specimen.

techniques [19, 20]. If the specimen is a dynamic object, the zero-order light can be removed by subtracting two holograms obtained sequentially [21–23].

6.2 Sectioning in holography

Figure 6.7 illustrates the imaging of a three-dimensional specimen by a simple objective lens. The objective lens focuses the object plane on the image plane sharply but on the image plane there are other contributions from out-of-focus planes above and below the "object plane." These out-of-focus contributions are usually termed out-of-focus haze in three-dimensional microscopy. Hence the image plane contains the in-focus as well as the out-of-focus information, and extraction of the in-focus information from the image plane, known as *sectioning* in three-dimensional microscopy, is required. Once a particular image plane is focused and consequently sectioned, the specimen can be moved up or down so that the other object planes within the specimen can be focused and sectioned in order to cover the whole three-dimensional specimen.

The advantage of holography is its ability to capture three-dimensional information from a large volume at high resolution. In contrast, the advantage of sectioning is the ability to section a thin image slice and eliminate contributions from other image sections within a three-dimensional volume.

These two advantages are seemingly incompatible. Recently, attempts have been made to reconcile the advantage of holography with that of sectioning. As discussed in Section 5.3, many designs using a low-coherence light source in standard digital holographic microscopes have a sectioning capability of about 10–20 μm, and sectioning capability with submicrometer accuracy has been demonstrated with optical scanning holography (OSH) but only with simple fluorescent beads [24] (see also Section 5.4 on optical scanning holography). To gain an appreciation of the type of work involved in sectioning, let us formulate the problem of sectioning in the context of optical scanning holography (OSH) as an example.

Let us assume that we have an object with complex amplitude $T(x, y; z)$. The complex Fresnel zone plate hologram obtained by OSH is given by [see Eq. (5.55)]

$$H_c(x,y) = \int |T(x,y;z)|^2 * \frac{k_0}{2\pi z} \exp\left[j\frac{k_0}{2z}(x^2+y^2)\right] dz, \tag{6.13}$$

which can be represented in a discrete form when we discretize the object into N sections at locations $z_1, z_2, \ldots, z_N$:

$$H_c(x,y) = \sum_{i=1}^{N} \left|T(x,y;z_i)\right|^2 * \frac{k_0}{2\pi z_i} \exp\left[j\frac{k_0}{2z_i}(x^2+y^2)\right]. \tag{6.14}$$

To reconstruct a section, say, at z_1, we simply convolve the complex hologram with a spatial impulse response in Fourier optics,

$$h(x,y;z) = \exp(-jk_0 z)\frac{jk_0}{2\pi z} \exp\left[\frac{-jk_0}{2z}(x^2+y^2)\right],$$

at $z = z_1$ since

$$\frac{k_0}{2\pi z_i} \exp\left[j\frac{k_0}{2z_i}(x^2+y^2)\right] * h(x,y;z=z_i) \propto \delta(x,y). \tag{6.15}$$

Hence the reconstruction at $z = z_1$ is given by

$$\begin{aligned} &H_c(x,y) * h(x,y;z_1) \\ &= \sum_{i=1}^{N} \left|T(x,y;z_i)\right|^2 * \frac{k_0}{2\pi z_i} \exp\left[j\frac{k_0}{2z_i}(x^2+y^2)\right] * h(x,y;z=z_1) \\ &= |T(x,y;z_1)|^2 + \sum_{i=2}^{N} \left|T(x,y;z_i)\right|^2 * \frac{k_0}{2\pi z_i} \exp\left[j\frac{k_0}{2z_i}(x^2+y^2)\right] * h(x,y;z=z_1), \end{aligned} \tag{6.16}$$

where we have used Eq. (6.15) to extract the section at $z = z_1$, which is the first term of the above equation. The second term in Eq. (6.16) is what has been referred to as the out-of-focus haze. The idea of sectioning in holography is to extract the first term from the hologram and at the same time to reject the second term.

Example 6.1: Demonstration of the out-of-focus haze

Although Eq. (6.14) is based on the principle of optical scanning holography (OSH), it can be regarded as a general form of complex hologram because it represents the summation of the object light diffracted from different planes. In this example, two rectangular objects (or two slits) at different depths are simulated. The setup is shown in Fig. 6.8, in which object 1 is a vertical slit $z_1 = 30$ cm from the hologram plane; object 2 is a horizontal slit $z_2 = 15$ cm from the hologram plane. The spatial domain is 512×512 pixels with pixel size of 10 μm, and the wavelength is 0.6328 μm. The complex hologram is generated according to Eq. (6.14), using the angular spectrum method. The simulation results are shown in Fig. 6.9, while the MATLAB code is listed in Table 6.1.

The real part and the imaginary part of the complex hologram are shown respectively in Fig. 6.9(a) and (b), which are also the two outputs, Out_1 and Out_2, of the optical holographic system in Fig. 5.14 [see Eq. (5.51)]. As we have the complex hologram, we can use Eq. (6.16) to focus (reconstruct) the hologram at any depth of interest. In Fig. 6.9(c) the focused depth is at $z = z_1$. We can see the vertical slit is sharply reconstructed but the horizontal slit is blurred. Similarly, when the focused depth is at $z = z_2$ in Fig. 6.9(d), the horizontal slit is sharp but the vertical slit is blurred. It should be noted that although the out-of-focus image is blurred, it is still apparent and thus may disturb the measurement of the in-focus portion.

Sectioning in digital holography is a relatively new topic. Research in this area seems to have started worldwide in 2006. Asundi and Singh from Singapore [25] proposed sectioning in-line digital holograms to display only the in-focus plane using information from a single out-of-focus plane. The method seems to work only for very fine particles. Leith's group from the USA [26] discussed optical sectioning in the context of optical scanning holography (OSH). Kim from Korea [27] proposed the use of Wiener filtering to reduce the defocus noise in the reconstruction of a three-dimensional image from a complex hologram obtained

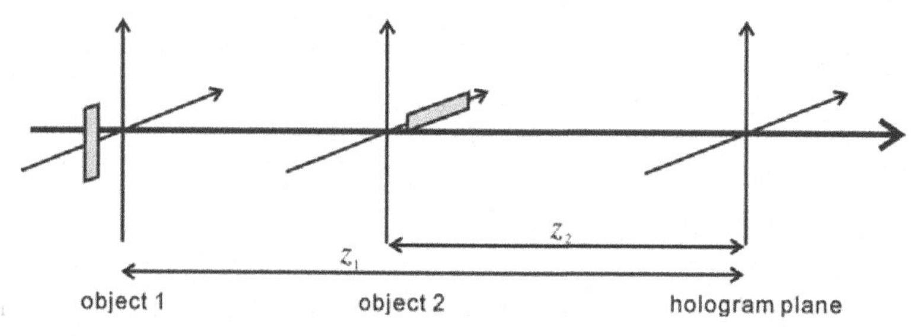

Figure 6.8 Geometry for a hologram of two rectangular objects at different depths.

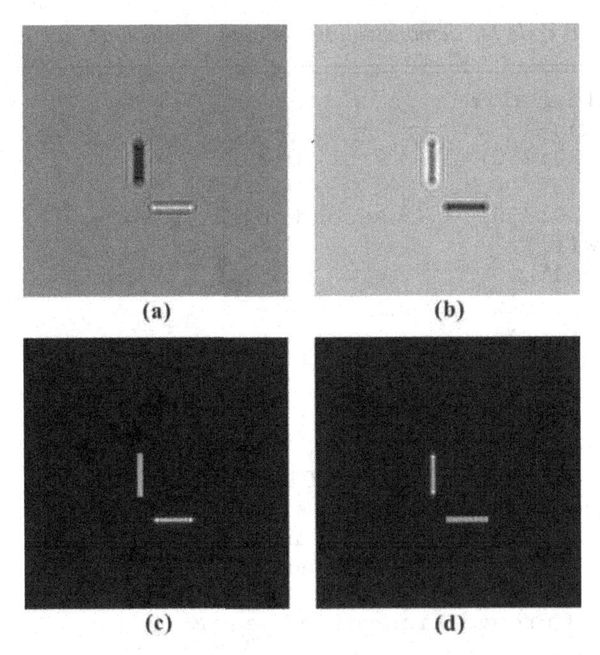

Figure 6.9 (a) Real part, and (b) imaginary part of the complex hologram of two rectangular objects, and the reconstructed images focused at (c) $z = z_1$, and (d) $z = z_2$.

by OSH. Iterative methods based on optimization using the L_2 norm [28] and filtering in the phase-space domain [29] to achieve optical sectioning have also been proposed in the context of digital holography.

The work of Zhang *et al.* deserves to be mentioned since it was the first reported use of the L_2 norm in holography to perform sectioning [28, 30]. Based on Eq. (6.16), Fig. 6.10 shows the reconstruction of a three-dimensional specimen consisting of a slide of fluorescent beads 2 μm in diameter (excitation around 542 nm, emission around 612 nm) for two sections. The beads tend to stick either to the top surface of the mounting slide or to the bottom surface of the coverslip, giving us a simple three-dimensional specimen with two dominant sections. The distance between the two sections is around 35 μm. Figure 6.10(a) and (b) show the section on the top of the slide and on the bottom of the slide, respectively. When the beads are not sharply focused on the section, we can see the haze from the out-of-focus plane. Figure 6.11(a) and (b) show the section on the top of the slide and on the bottom of the slide, respectively, when the L_2 norm is employed in the reconstruction. We can clearly see that the haze has been eliminated on both of the planes.

The first use of the L_2 norm in digital holography seems to have opened up vibrant research into performing sectioning directly from a hologram using other optimization methods such as the well-known L_1 norm, already widely used in signal processing, which ties to what is often nowadays called *compressive holography* [31]. The use of compressive holography has allowed, for example,

Table 6.1 *MATLAB code for demonstrating the out-of-focus haze, see Example 6.1*

```
clear all; close all;
I1=zeros(512);
I1(220:300,220:230)=1; % first object
I2=zeros(512);
I2(340:350,250:330)=1; % second object
figure;imshow(I1);
title('Object 1')
axis off
figure;imshow(I2);
title('Object 2')
axis off
% parameter setup
M=512;
deltax=0.001; % pixel pitch 0.001 cm (10 um)
w=633*10^-8; % wavelength 633 nm
z1=30; % 25 cm, propagation distance z1
z2=15; % 15 cm, propagation distance z2

%Simulation of propagation using the ASM
r=1:M;
c=1:M;
[C, R]=meshgrid(c, r);
deltaf=1/M/deltax;
A01=fftshift(ifft2(fftshift(I1)));
p1=exp(-2i*pi*z1.*((1/w)^2-((R-M/2-1).*deltaf).^2-...
    ((C-M/2-1).*deltaf).^2).^0.5);
Az1=A01.*p1;
A02=fftshift(ifft2(fftshift(I2)));
p2=exp(-2i*pi*z2.*((1/w)^2-((R-M/2-1).*deltaf).^2-...
    ((C-M/2-1).*deltaf).^2).^0.5);
Az2=A02.*p2;
Hologram=fftshift(fft2(fftshift(Az1)))+...
   fftshift(fft2(fftshift(Az2)));
Hr=mat2gray(real(Hologram));%Real part of the hologram
Hi=mat2gray(imag(Hologram));%Imaginary part

figure;imshow (Hr)
title('Real part of the complex hologram')
axis off
figure;imshow (Hi)
title('Imaginary part of the complex hologram')
axis off

%Reconstruction
Ar=fftshift(ifft2(fftshift(Hologram)));
Arz1=Ar.*conj(p1);
```

Table 6.1 (*cont.*)

```
EI1=fftshift(fft2(fftshift(Arz1)));
EI1=mat2gray(EI1.*conj(EI1));
figure; imshow(EI1);
title('Reconstructed image at z=z_1')
axis off
Arz2=Ar.*conj(p2);
EI2=fftshift(fft2(fftshift(Arz2)));
EI2=mat2gray(EI2.*conj(EI2));
figure; imshow(EI2);
title('Reconstructed image at z=z_2')
axis off
```

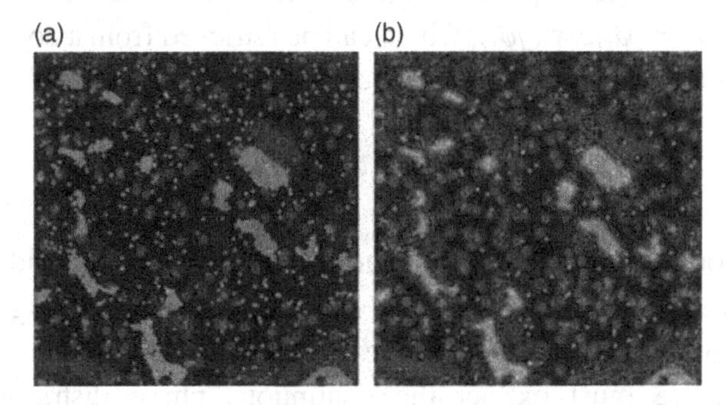

Figure 6.10 Sectioning by conventional reconstruction. (a) Top section of the specimen, and (b) bottom section of the specimen. From [Ref. 30], with permission, © OSA.

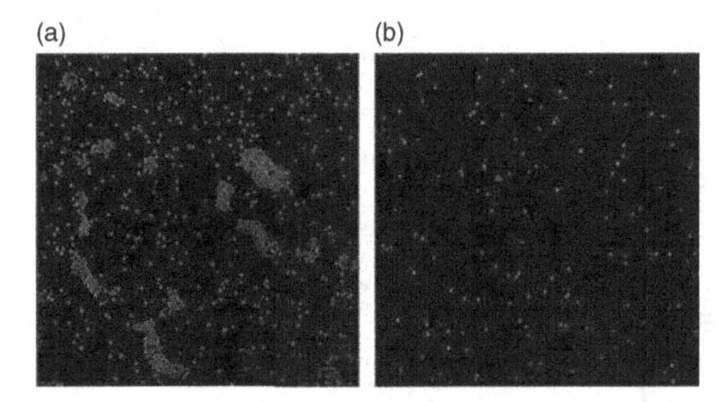

Figure 6.11 Sectioning by the L_2 norm. (a) Top section of the specimen, and (b) bottom section of the specimen. From [Ref. 30], with permission, © OSA.

rejection of the twin image noise as well as out-of-focus haze for on-axis holograms. Most recently, Zhao *et al.* have proposed an adaptively iterative shrinkage-thresholding algorithm to show that the performance of the algorithm in terms of sectioning capability is better than that of the L_2 norm method [32]. In addition, the proposed algorithm can section well even when using only half of the data of a complex hologram. It is expected that compressive holography will continue to flourish, for example, with better and faster algorithms, for application to sectioning and to different challenges in holography.

6.3 Phase extraction

One of the most important merits of digital holography is that the phase of the sample can be determined quantitatively. This may find important applications in microscopy for biological specimens. We assume that the complex field at the sample plane is $\psi_s = |\psi_s|\exp(j\phi_s)$, which can be retrieved from a complex hologram obtained by digital holography. Consequently, the phase can be obtained by

$$\phi_w = \tan^{-1}\left\{\frac{\mathrm{Re}\{\psi_s\}}{\mathrm{Im}\{\psi_s\}}\right\}. \tag{6.17}$$

However, the output range of the arctangent is between $-\pi$ and π. The phase obtained from Eq. (6.17) is thus wrapped to $-\pi \sim \pi$, as shown in Fig. 6.12. So even if we have ϕ_w, the actual phase ϕ_s can be $\phi_w + 2\pi l$ for any integer l. To eliminate this ambiguity, we must extract the continuous phase distribution from the wrapped phase. This procedure is called *phase-unwrapping* or *phase-tracking*.

A straightforward method for phase-unwrapping is to detect and compensate for the phase jump. Here we consider a one-dimensional case as an example. First we calculate the phase difference between adjacent pixels,

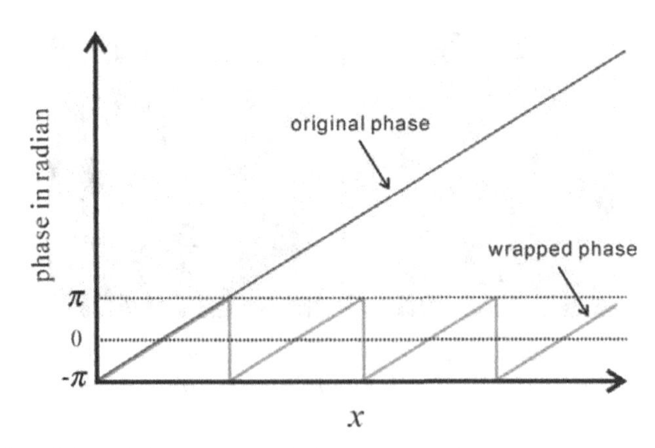

Figure 6.12 The original phase and the wrapped phase.

$$\Delta[n] = \phi_w[n] - \phi_w[n-1], \tag{6.18}$$

where $n = 1 \sim N$ is the pixel index. If $\Delta[n]$ is larger than $+\pi$ or smaller than $-\pi$, then the phase jumps to the lower or upper 2π range, respectively. We must, therefore, add -2π or 2π to the phase of all the following pixels. If none of the above conditions is valid, the phase value remains unchanged. This procedure can be done by introducing an intermediate function, $g[n]$,

$$g[n] = \begin{cases} g[n-1] - 2\pi & \text{for } \Delta[n] > \pi \\ g[n-1] + 2\pi & \text{for } \Delta[n] < -\pi \\ g[n-1] & \text{otherwise.} \end{cases} \tag{6.19}$$

By setting the initial conditions $\phi_{uw}[1] = \phi_w[1]$, $\Delta[1] = 0$, and $g[1] = 0$, the unwrapped phase $\phi_{uw}[n]$ can be obtained sequentially by

$$\phi_{uw}[n] = \phi_w[n] + g[n]. \tag{6.20}$$

It has also been indicated that Eqs. (6.19) and (6.20) can be expressed using a single formula as [33]

$$\phi_{uw}[n] = \phi_w[1] + \sum_{m=1}^{n-1} \mathrm{W}\{\Delta[m]\}, \tag{6.21}$$

where $W\{\bullet\}$ is the wrapping operator. The wrapping operator can be regarded as a non-linear calculation as

$$\mathrm{W}\{\Delta[m]\} = \Delta[m] + 2\pi l[m], \tag{6.22}$$

where $l[m] \in \{\ldots, -1, 0, 1, \ldots\}$, i.e., it is a sequence of integers chosen to ensure that the wrapped function, $\mathrm{W}\{\Delta[m]\}$, is between $-\pi$ and π. The last term in Eq. (6.21) represents a summation of the wrapped phase difference $\Delta[n]$. Equation (6.21) is usually called the *Itoh algorithm*.

Example 6.2: Phase unwrapping of a wrapped phase map

In this example, a phase function

$$\phi_s(x, y) = 15 \times \sin\left[\frac{(x-225)^2 + (y-200)^2}{1000}\right] + \left[\frac{(x-100)^2 + (y-37)^2}{500}\right]$$

is first produced in the range $x = 0 \sim 256$, $y = 0 \sim 256$, and wrapped. The sampling separation is 0.5, and so the wrapped phase function contains 513×513 pixels. The wrapped phase is then unwrapped using the Itoh algorithm.

In this example, we do not write the MATLAB code for implementing Eq. (6.21), rather we use directly the MATLAB command `unwrap` to unwrap the

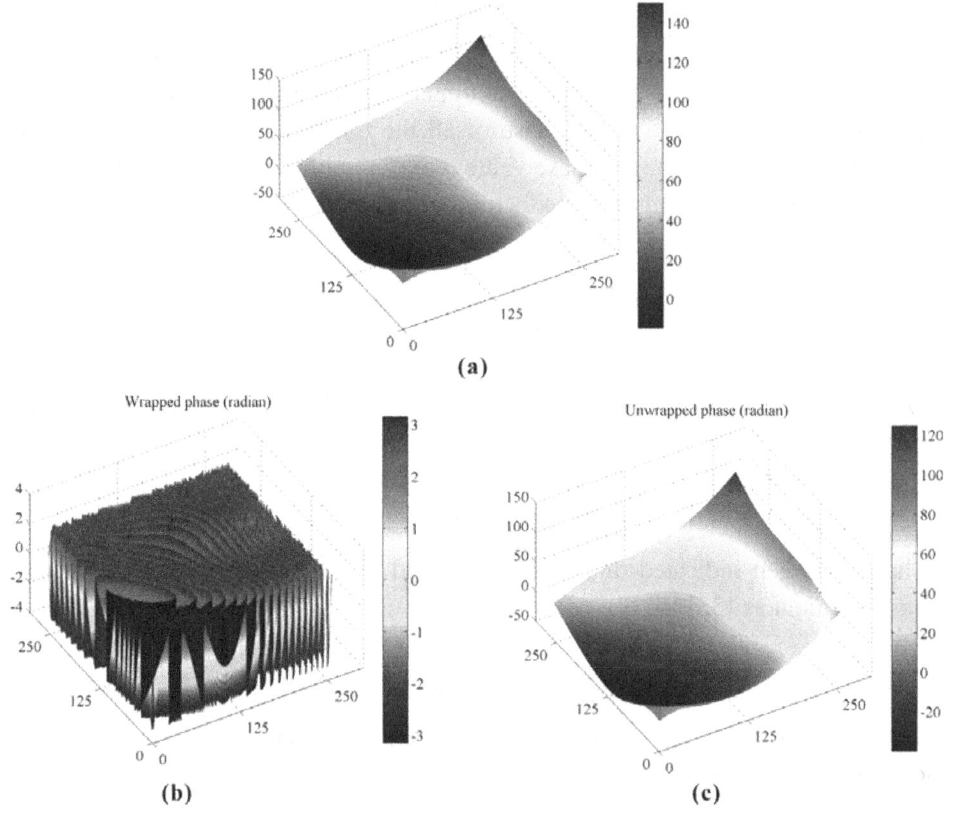

Figure 6.13 (a) The original phase, (b) the wrapped phase, and (c) the unwrapped phase.

wrapped phase function. It should be noted, however, that unwrap only applies to a single dimension. So we must apply unwrap twice, once for the columns and once for the rows. The simulation results are shown in Fig. 6.13, and the MATLAB code is listed in Table 6.2. Figure 6.13(a) shows the original phase ranging from −15 to 150 radians. The wrapped phase, shown in Fig. 6.13(b), ranges from −3.14 to 3.14 radians. The unwrapped phase is shown in Fig. 6.13(c), which is nearly the same as the phase shown in Fig. 6.13(a). Because the phase ϕ_{uw} is unwrapped from some start point, there is a difference at the start point, $\phi_{uw}[n] - \phi_s[n]$, between the unwrapped phase ϕ_{uw} and the exact phase ϕ_s. In the current example of a two-dimensional phase function, we find the phase difference $\phi_{uw}[1,1] - \phi_s[1,1] = 2.92 - 28.05 = -25.13$ radians, where we have used the start point at coordinates $(x, y) = (0,0)$ to find the phase difference. Indeed $\phi_{uw}[1,1]$ is the first point when $(x, y) = (0,0)$. We can also use other start points to wrap the phase function but for this we need to write a code instead of using the MATLAB command. So in

Table 6.2 *MATLAB code for two-dimensional phase-unwrapping, see Example 6.2*

```
clear all; close all;
[x,y]=meshgrid(0:0.5:256,0:0.5:256);
PHASE=15*sin(((x-200).^2+(y-225).^2)/10000)...
+0.002*((x-37).^2+(y-100).^2);
% original phase function
FUN=exp(1i*PHASE);
WRAP_PHASE=angle(FUN);% wrapped phase function
UNWRAP_PHASE=unwrap(WRAP_PHASE,[ ],1);%unwrapping columns
UNWRAP_PHASE=unwrap(UNWRAP_PHASE,[ ],2);%unwrapping rows

figure; mesh(x,y,PHASE);view([-30,52])
title('Original phase (radian)')
colorbar
set (gca, 'xtick', [0 125 250]);
set (gca, 'ytick', [0 125 250]);
figure; mesh(x,y,WRAP_PHASE);view([-30,52])
title('Wrapped phase (radian)')
colorbar
set (gca, 'xtick', [0 125 250]);
set (gca, 'ytick', [0 125 250]);
figure; mesh(x,y,UNWRAP_PHASE);view([-30,52])
title('Unwrapped phase (radian)')
colorbar
set (gca, 'xtick', [0 125 250]);
set (gca, 'ytick', [0 125 250]);
```

Fig. 6.13(c) the unwrapped phase ranges roughly from –40 to 125 radians. The added constant phase difference of –25.13 radians in Fig. 6.13(c) compared with the original phase in Fig. 6.13(a) does not matter because only the relative phase is meaningful.

Although in Example 6.2 we have demonstrated unwrapping of a two-dimensional phase function successfully and easily, there are some problems with the Itoh algorithm. First, Eq. (6.21), as well as Eq. (6.19), implies that the phase difference between adjacent pixels of ϕ_s must be between $-\pi$ and π. If the exact phase changes abruptly (say, larger than π), the algorithm fails and the unwrapped phase ϕ_{uw} suffers from errors.

For example, we assume that $\phi_s[1] = 0.9\pi$ and $\phi_s[2] = 2.1\pi$, where the phase difference is not between $-\pi$ and π. The wrapped phases of the two pixels are then $\phi_w[1] = 0.9\pi$ as there is no need to wrap the phase, and $\phi_w[2] = 2.1\pi - 2\pi = 0.1\pi$ (l has been chosen to be -1) so that the wrapped phase is in the range between $-\pi$ and π. According to Eq. (6.18), $\Delta[2] = \phi_w[2] - \phi_w[1] = -0.8\pi$ with initial assumed

$\Delta[1] = 0$. Hence according to Eq. (6.19), $g[2] = g[1] = 0$. We can now, from Eq. (6.20), find $\phi_{uw}[1] = \phi_w[1] + g[1] = 0.9\pi$, which is correct. But $\phi_{uw}[2] = \phi_w[2] + g[2] = 0.1\pi$, which is incorrect.

In addition, in practice the actual acquired phase image is usually contaminated with noise. Any noisy pixel may affect the judgment described in Eq. (6.19), and the phase may be shifted incorrectly up or down. What is worse is that incorrect determination of the phase on the nth pixel will propagate to the following ($n + 1$, $n + 2, \ldots$) pixels. Many ingenious, noise-immune unwrapping methods have been proposed. For example, the noisy pixels are usually detected and excluded from the unwrapping procedure (e.g., [34, 35]).

6.4 Optical contouring and deformation measurement

Because of its fast, non-contact, and three-dimensional imaging ability, digital holography can be applied to profile measurements or optical contouring. We assume that the digital hologram is taken using the on-axis, reflection-type digital holography, as shown in Fig. 6.2. For simplicity, we also assume that complex holograms are obtained by phase shifting technique. The complex holograms are then reconstructed and unwrapped, and finally the continuous phase on the specimen is extracted. In Fig. 6.14, we only show the object beam of the reflection-type microscope-based digital holographic microscope of Fig. 6.2 for the following analysis. We consider a plane wave emitting from reference plane A, which illuminates the specimen and retro-reflects back to plane A. The phase delay of the light is proportional to the optical path, and thus the unwrapped phase can be expressed as

$$\phi_{uw}[n] = (-2k_0 d) - (-2k_0 h[n]), \tag{6.23}$$

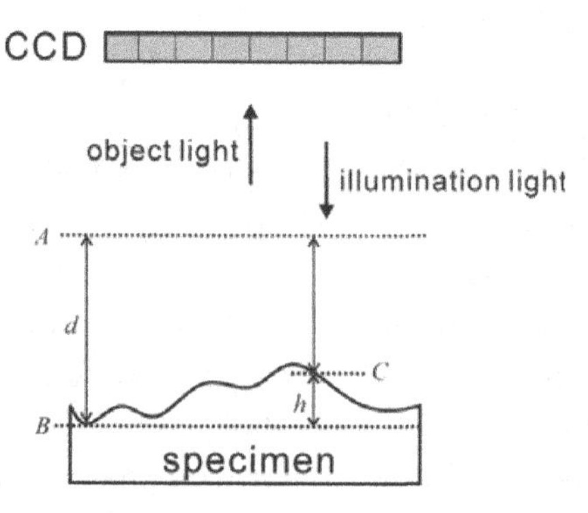

Figure 6.14 Optical contouring geometry.

where d is the separation between plane A and another reference plane, plane B; h is the height of the specimen measured from plane B. Therefore, the height of the specimen is given by

$$h[n] = \frac{\phi_{uw}[n] + 2k_0 d}{2k_0} = \frac{\phi^b_{uw}[n]}{2k_0}, \tag{6.24}$$

where $\phi^b_{uw}[n]$ is the unwrapped phase found by setting a zero reference phase at plane B. Thus the height separation between 2π phase contouring is

$$\Delta h_\lambda = \frac{2\pi}{2k_0} = \frac{\lambda_0}{2}, \tag{6.25}$$

which is also called the *height sensitivity* of the system. Hence we can see that each *contour* is the locus of all points on the surface of the specimen that lie at some constant height above a reference plane.

Whenever the height difference of the specimen is less than $\Delta h_\lambda/2 = \lambda_0/4$, we have $\phi^b_{uw}[n] = \phi_s$ as the phase difference between adjacent pixels is always less than π. In other words, unwrapping is not necessary if the height variation of a specimen is less than $\lambda_0/4$. However, such height sensitivity in the system being considered is rather small and is sometimes impractical. Hence there is a need to have a system of large height sensitivity. Two-wavelength contouring, to be discussed next, is a solution.

6.4.1 Two-wavelength contouring

To extend the range of height measurement without the ambiguity in the procedure of phase-unwrapping, one can use multiple light sources with different wavelengths to acquire digital holograms sequentially [36–38]. Here we take *two-wavelength contouring* as an example. We assume that the continuous phase functions corresponding to the two wavelengths λ_1 and λ_2 ($\lambda_1 > \lambda_2$) are ϕ^1_s and ϕ^2_s, respectively, where

$$\phi^1_s[n] = [-2k_1 d] + 2k_1 h[n], \tag{6.26a}$$

$$\phi^2_s[n] = [-2k_2 d] + 2k_2 h[n]. \tag{6.26b}$$

The phase difference is

$$\Delta\phi_s[n] = \phi^2_s[n] - \phi^1_s[n] = 2(k_2 - k_1)h[n] + 2d(k_1 - k_2). \tag{6.27}$$

On the right hand side of Eq. (6.27), the second term is a constant phase, and the first term can be regarded as the phase delay of light with a *synthetic wavelength* or *equivalent wavelength*

$$\Lambda = \frac{\lambda_1\lambda_2}{\lambda_1-\lambda_2}. \tag{6.28}$$

By setting $\lambda_1 - \lambda_2 = \Delta\lambda \ll \lambda_1$, the synthetic wavelength, $\Lambda \approx \lambda_1^2/\Delta\lambda$, is much larger than λ_1. And, according to Eq. (6.25), the height sensitivity is

$$\Delta h_\Lambda = \frac{\Lambda}{2} \approx \frac{\lambda_1^2}{2\Delta\lambda}. \tag{6.29}$$

For example, supposing that $\lambda_1 = 640.2$ nm and $\lambda_2 = 640$ nm, we will have $\Delta h_\lambda \approx$ 320 nm and $\Delta h_\Lambda \approx 1024$ μm. Thus the measurement range of height is significantly extended without the need to perform unwrapping.

If unwrapping is needed, the two-wavelength procedure begins with two wrapped phase functions obtained by digital holography,

$$\phi_w^1[n] = W\{\phi_s^1[n]\} = \phi_s^1[n] + 2\pi l_1[n], \tag{6.30a}$$

$$\phi_w^2[n] = W\{\phi_s^2[n]\} = \phi_s^2[n] + 2\pi l_2[n], \tag{6.30b}$$

where l_1 and l_2 are integers to ensure ϕ_w^1 and ϕ_w^2 are in the $\pm\pi$ range, respectively. Note that again according to the Itoh algorithm, ϕ_w^1 or ϕ_w^2 cannot be unwrapped to a correct unwrapped phase function provided the phase jump between adjacent pixels of ϕ_s^1 or ϕ_s^2 is larger than π. The difference between the two wrapped functions is

$$\begin{aligned} \Delta\phi_w[n] &= \phi_w^2[n]-\phi_w^1[n] \\ &= \phi_s^2[n]-\phi_s^1[n] + 2\pi\left[l_2[n]-l_1[n]\right] \\ &= \Delta\phi_s[n] + 2\pi\left[l_2[n]-l_1[n]\right]. \end{aligned} \tag{6.31}$$

Consequently, by unwrapping $\Delta\phi_w[n]$, a continuous function $\Delta\phi_s[n]$ can be obtained. Similar to Eq. (6.24), the final profile of the specimen can be given by

$$h[n] = \frac{\Delta\phi_s(n)}{2\times 2\pi/\Lambda} = \frac{\Delta\phi_s(n)}{4\pi}\Lambda. \tag{6.32}$$

Although two-wavelength contouring can efficiently extend the measurement range of height, there are two remaining problems. First, the method cannot correctly measure structures with a sharp height difference of more than Λ/4 because a phase jump of more than $\pm\pi$ occurs at the edges of the element. The other problem is that the height error due to the phase error will be amplified in the subtraction process [see Eq. (6.31)]. The two problems can be eliminated together by involving a third or more wavelengths [37, 39–41], i.e., multi-wavelength contouring.

Example 6.3: One-dimensional example of two-wavelength optical contouring

In this example, we would like to demonstrate two-wavelength optical contouring in a one-dimensional case. The simulation results are shown in Fig. 6.15, and

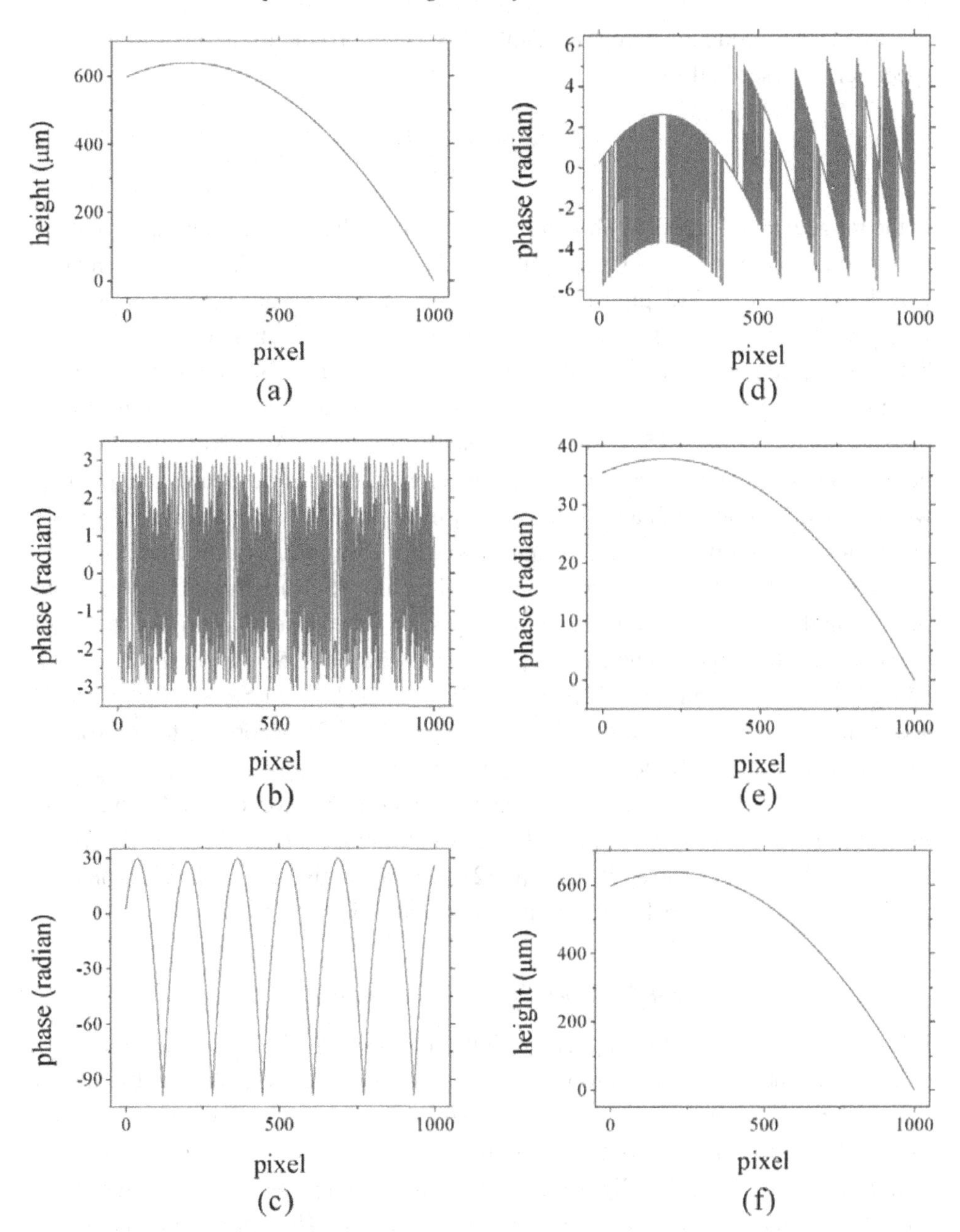

Figure 6.15 Two-wavelength optical contouring. (a) Specimen profile; (b) wrapped phase for a single wavelength; (c) unwrapped phase from (b); (d) wrapped phase for a synthetic wavelength; (e) unwrapped phase from (d); (f) recovered profile.

the MATLAB code is given in Table 6.3. The profile of a one-dimensional specimen is described by

$$h[n] = 700-\frac{(n-201)^2}{1000}\ \mu\text{m}$$

in the range $n = 1 \sim 1000$, as shown in Fig. 6.15(a). The two wavelengths involved in the example are $\lambda_1 = 0.650$ μm and $\lambda_2 = 0.652$ μm. Thus the synthetic wavelength is $\Lambda = 212$ μm, according to Eq. (6.28). Figure 6.15(b) shows the wrapped phase for λ_1. It looks tumultuous because the actual phase difference between adjacent pixels exceeds the $\pm\pi$ limit. As a result, using the Itoh algorithm, the unwrapped phase shown in Fig. 6.15(c) cannot correctly represent the profile of the specimen. Figure 6.15(d) shows the phase difference $\Delta\phi_w$. Note that because both ϕ_w^1 and ϕ_w^2 range between $\pm\pi$, their difference, $\Delta\phi_w$, is between $\pm 2\pi$. In other words, the difference of l_2 and l_1, i.e., $l_2 - l_1$ in Eq. (6.31), is not always the proper parameter for wrapping $\Delta\phi_s$ to the $\pm\pi$ range. For example, assume that $\Delta\phi_s[1] = 0.2\pi$ and $\Delta\phi_s[2] = 0.5\pi$. The phase obtained by Eq. (6.31) is $\Delta\phi_w[1] = 0.2\pi$, but $\Delta\phi_w[2]$ might have wrapped to the range beyond $\pm\pi$ because of improper choice of the parameter $(l_2 - l_1)$, and hence $\Delta\phi_w[2] = 0.5\pi - 2\pi = -1.5\pi$. By setting $\Delta[1] = 0$, and $g[1] = 0$, we can get $\Delta[2] = \Delta\phi_w[2] - \Delta\phi_w[1] = -1.5\pi - 0.2\pi = -1.7\pi$ according to Eq. (6.18), and thus $g[2] = g[1] + 2\pi = 2\pi$ according to Eq. (6.19). Finally, from Eq. (6.20) the unwrapped phase will be $\Delta\phi_{uw}[2] = \Delta\phi_w[2] + g[2] = -1.5\pi + 2\pi = 0.5\pi$, which is the correct phase. So we can still apply the Itoh algorithm to unwrap $\Delta\phi_w$, yielding a continuous phase $\Delta\phi_s$, as shown in Fig. 6.15(e). Finally, we apply Eq. (6.32) to convert the phase distribution to a profile function in Fig. 6.15(f), which is identical to Fig. 6.15(a).

6.4.2 Two-illumination contouring

If for some reason we cannot apply two light sources in the holographic system, we can employ the two-illumination contouring method [42, 43]. Again, the object arm of the holographic system is highlighted in Fig. 6.16. The geometry is similar to that shown in Fig. 6.14 but now the plane-wave illumination is tilted from the normal at an angle θ. The light emitting from reference plane A illuminates the specimen (plane C) and reflects back to plane A again. The phase delay is

$$\phi_s^1 = [-k_0 \sin\theta x - k_0 \cos\theta(d-h)] + [-k_0(d-h)], \tag{6.33}$$

where the first two terms correspond to the optical path from plane A to plane C, and the third term corresponds to the optical path from plane C to plane A. In the second recording, the angle of illumination light is changed to $\theta + \Delta\theta$. The phase delay becomes

Table 6.3 *MATLAB code for demonstrating two-wavelength optical contouring, see Example 6.3*

```
clear all; close all;
x=1:1:1000;
h=ones(1,1000);
h=700-h.*((x-201).^2)/1000;
h=h-min(h); % Sample profile in um
L1=0.650;   % first wavelength in um
L2=0.652;   % second wavelength in um
Ls=L1*L2/(L2-L1); % synthetic wave length in um

%Phase function for the two wavelengths
PHASE1=4*pi*h/L1-4*pi*700/L1;
PHASE2=4*pi*h/L2-4*pi*700/L2;
FUN1=exp(1i*PHASE1);
WRAP_PHASE1=angle(FUN1);% wrapped phase function 1
FUN2=exp(1i*PHASE2);
WRAP_PHASE2=angle(FUN2);% wrapped phase function 2

% unwrapping phase function 1
UNWRAP_PHASE1=unwrap(WRAP_PHASE1,[ ],2);

% difference between two wrapped phase functions
PHASE3=WRAP_PHASE1-WRAP_PHASE2;

UNWRAP_PHASE3=unwrap(PHASE3,[ ],2);
% unwrapping synthetic phase function

UNWRAP_PHASE3=UNWRAP_PHASE3-min(UNWRAP_PHASE3);
H=UNWRAP_PHASE3*Ls/4/pi;% corresponding height
figure; plot (h);
title('Sample profile')
xlabel('pixel');ylabel('height (\mum)')
figure; plot (WRAP_PHASE1);
title('Wrapped phase 1')
xlabel('pixel');ylabel('Phase (radian)')
figure; plot (UNWRAP_PHASE1);
title('Unwrapped phase 1')
xlabel('pixel');ylabel('Phase (radian)')
figure; plot (PHASE3);
title('Wrapped phase for synthetic wavelength')
xlabel('pixel');ylabel('Phase (radian)')
figure; plot (UNWRAP_PHASE3);
title('Unwrapped phase for synthetic wavelength')
xlabel('pixel');ylabel('Phase (radian)')
figure; plot (H);
title('Recovered profile of the sample')
xlabel('pixel');ylabel('height (\mum)')
```

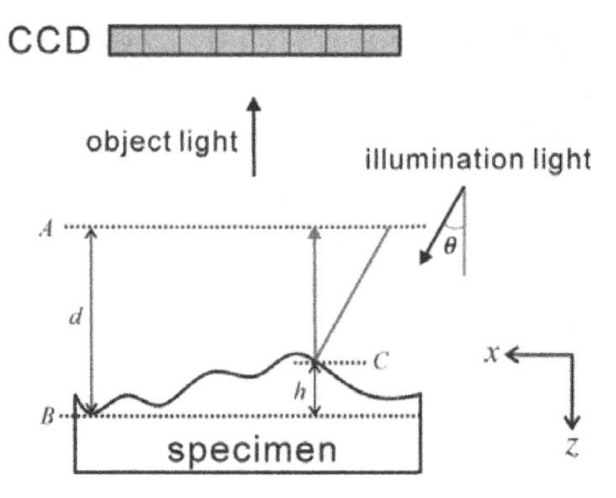

Figure 6.16 Two-illumination contouring geometry.

$$\phi_s^2 = [-k_0 \sin(\theta + \Delta\theta)x - k_0 \cos(\theta + \Delta\theta)(d-h)] + [-k_0(d-h)]. \tag{6.34}$$

The difference of the two phase functions is thus

$$\Delta\phi_s = \phi_s^2 - \phi_s^1 = 2k_0 \sin\frac{\Delta\theta}{2}\left[\sin\left(\theta + \frac{\Delta\theta}{2}\right)(d-h) - \cos\left(\theta + \frac{\Delta\theta}{2}\right)x\right] \tag{6.35}$$

by applying two identities

$$\sin\alpha - \sin\beta = 2\cos\left(\frac{\alpha+\beta}{2}\right)\cos\left(\frac{\alpha-\beta}{2}\right),$$

$$\cos\alpha - \cos\beta = -2\sin\left(\frac{\alpha+\beta}{2}\right)\sin\left(\frac{\alpha-\beta}{2}\right).$$

The x-dependent term in Eq. (6.35) is a familiar carrier term as in a carrier-frequency hologram [see Eq. (3.5b)] and its phase term (the first term of the right hand side of the equation) can be extracted to yield a height-dependent phase as

$$\Delta\phi_s' = 2k_0 \sin\frac{\Delta\theta}{2}\sin\left(\theta + \frac{\Delta\theta}{2}\right)(d-h). \tag{6.36}$$

Therefore, the specimen height can be retrieved by

$$h = d - \frac{\Delta\phi_s'}{2k_0 \sin\frac{\Delta\theta}{2}\sin\left(\theta + \frac{\Delta\theta}{2}\right)}. \tag{6.37}$$

Accordingly, the height sensitivity corresponding to 2π contouring is

$$\Delta h = \frac{\lambda_0}{2\sin\frac{\Delta\theta}{2}\sin\left(\theta + \frac{\Delta\theta}{2}\right)} \approx \frac{\lambda_0}{\Delta\theta \sin\theta} \tag{6.38}$$

for a small tilt angle $\Delta\theta$. For example, for $\lambda_0 = 640$ nm, $\theta = 20°$, and $\Delta\theta = 0.05°$, the height sensitivity is $\Delta h = 2144$ μm.

6.4.3 Deformation measurement

Sometimes, such as measurement of the Young modulus of a material, it is the profile deformation instead of the whole profile which is of interest. Deformation measurement can also be realized using the setup shown in Fig. 6.16 [44]. First a hologram corresponding to the original profile is recorded. The retrieved phase function is described by Eq. (6.33). The specimen undergoes a small deformation, and a second hologram is recorded. The retrieved phase function of the second hologram is

$$\phi_s^2 = [-k_0 \sin\theta x - k_0 \cos\theta(d-h-\delta h)] + [-k_0(d-h-\delta h)], \tag{6.39}$$

where δh denotes the height deformation of the specimen. As a result, the difference between the first and second phase functions is

$$\Delta\phi_s = \phi_s^2 - \phi_s^1 = k_0(1 + \cos\theta)\delta h. \tag{6.40}$$

The height deformation is thus given by

$$\delta h = \frac{\Delta\phi_s}{k_0(1 + \cos\theta)}. \tag{6.41}$$

In two-illumination contouring, some portions of the specimen may be shadowed because the illumination light is tilted. This problem together with any additive noise can be eliminated by applying more than two illuminations with different illumination directions [45, 46].

Problems

6.1 A CCD imager contains 1000 × 1000 pixels with a pixel pitch of 6 μm, and it is used in spherical-reference-based DHM. Find a combination of (z_0, z_r), where the distances z_0 and z_r are defined in Fig. 6.5(a), so that the acquired hologram is well sampled and the lateral resolution in the object space is better than 1.2 μm. Assume that phase-shifting procedures have been applied to extract a complex hologram and the wavelength used is $\lambda_0 = 0.6$ μm.

6.2 The CCD described in Problem 6.1 is applied in Fourier-based DHM. Find the object distance z_0 and the size of reconstructed field when a lateral resolution better than 1.2 μm is required. Assume that phase-shifting procedures have been applied to extract a complex hologram and the wavelength used is $\lambda_0 = 0.6$ μm.

References

1. D. Garbor, A new microscopic principle, *Nature* **161**, 777–778 (1948).
2. E. N. Leith, and J. Upatnieks, Microscopy by wavefront reconstruction, *Journal of the Optical Society of America* **55**, 569 (1965).

3. T. Zhang, and I. Yamaguchi, Three-dimensional microscopy with phase-shifting digital holography, *Optics Letters* **23**, 1221–1223 (1998).
4. E. Cuche, P. Marquet, and C. Depeursinge, Simultaneous amplitude-contrast and quantitative phase-contrast microscopy by numerical reconstruction of Fresnel off-axis holograms, *Applied Optics* **38**, 6994–7001 (1999).
5. P. Ferraro, S. De Nicola, A. Finizio, G. Coppola, S. Grilli, C. Magro, and G. Pierattini, Compensation of the inherent wave front curvature in digital holographic coherent microscopy for quantitative phase-contrast imaging, *Applied Optics* **42**, 1938–1946 (2003).
6. T. Colomb, E. Cuche, F. Charrière, J. Kühn, N. Aspert, F. Montfort, P. Marquet, and C. Depeursinge, Automatic procedure for aberration compensation in digital holographic microscopy and applications to specimen shape compensation, *Applied Optics* **45**, 851–863 (2006).
7. K. J. Chalut, W. J. Brown, and A. Wax, Quantitative phase microscopy with asynchronous digital holography, *Optics Express* **15**, 3047–3052 (2007).
8. J. W. Kang, and C. K. Hong, Phase-contrast microscopy by in-line phase-shifting digital holography: shape measurement of a titanium pattern with nanometer axial resolution, *Optical Engineering* **46**, 040506 (2007).
9. W. S. Haddad, D. Cullen, J. C. Solem, J. W. Longworth, A. McPherson, K. Boyer, and C. K. Rhodes, Fourier-transform holographic microscope, *Applied Optics* **31**, 4973–4978 (1992).
10. C. Wagner, S. Seebacher, W. Osten, and W. Jüptner, Digital recording and numerical reconstruction of lensless Fourier holograms in optical metrology, *Applied Optics* **38**, 4812–4820 (1999).
11. D. Dirksen, H. Droste, B. Kemper, H. Deleré, M. Deiwick, H. H. Scheld, and G. von Bally, Lensless Fourier holography for digital holographic interferometry on biological samples, *Optics and Lasers in Engineering* **36**, 241–249 (2001).
12. G. Pedrini, and H. J. Tiziani, Short-coherence digital microscopy by use of a lensless holographic imaging system, *Applied Optics* **41**, 4489–4496 (2002).
13. M. Sebesta, and M. Gustafsson, Object characterization with refractometric digital-Fourier holography, *Optics Letters* **30**, 471–473 (2005).
14. L. Granero, V. Micó, Z. Zalevsky, and J. García, Superresolution imaging method using phase-shifting digital lensless Fourier holography, *Optics Express* **17**, 15008–15022 (2009).
15. I. Yamaguchi, J.-I. Kato, S. Ohta, and J. Mizuno, Image formation in phase-shifting digital holography and applications to microscopy, *Applied Optics* **40**, 6177–6186 (2001).
16. J. W. Goodman, *Introduction to Fourier Optics* (McGraw-Hill, New York, 2005), pp. 108–115.
17. T. Colomb, J. Kühn, F. Charrière, C. Depeursinge, P. Marquet, and N. Aspert, Total aberrations compensation in digital holographic microscopy with a reference conjugated hologram, *Optics Express* **14**, 4300–4306 (2006).
18. J. Di, J. Zhao, W. Sun, H. Jiang, and X. Yan, Phase aberration compensation of digital holographic microscopy based on least squares surface fitting, *Optics Communications* **282**, 3873–3877 (2009).
19. V. Micó, J. García, Z. Zalevsky, and B. Javidi, Phase-shifting Gabor holography, *Optics Letters* **34**, 1492–1494 (2009).
20. V. Micó, L. Granero, Z. Zalevsky, and J. García, Superresolved phase-shifting Gabor holography by CCD shift, *Journal of Optics A, Pure and Applied Optics* **11**, 125408 (2009).

21. W. Xu, M. H. Jericho, H. J. Kreuzer, and I. A. Meinertzhagen, Tracking particles in four dimensions with in-line holographic microscopy, *Optics Letters* **28**, 164–166 (2003).
22. J. Garcia-Sucerquia, W. Xu, S. K. Jericho, P. Klages, M. H. Jericho, and H. J. Kreuzer, Digital in-line holographic microscopy, *Applied Optics* **45**, 836–850 (2006).
23. W. Xu, M. H. Jericho, I. A. Meinertzhagen, and H. J. Kreuzer, Digital in-line holography of microspheres, *Applied Optics* **41**, 5367–5375 (2002).
24. G. Indebetouw, A posteriori quasi-sectioning of the three-dimensional reconstructions of scanning holographic microscopy, *Journal of the Optical Society of America* **23**, 2657–2661 (2006).
25. A. Anand, and S. Vijay Raj, Sectioning of amplitude images in digital holography, *Measurement Science and Technology* **17**, 75–78 (2006).
26. W.-C. Chien, D. S. Dilworth, E. Liu, and E. N. Leith, Synthetic-aperture chirp confocal imaging, *Applied Optics* **45**, 501–510 (2006).
27. T. Kim, Optical sectioning by optical scanning holography and a Wiener filter, *Applied Optics* **45**, 872–879 (2006).
28. X. Zhang, E. Y. Lam, and T.-C. Poon, Reconstruction of sectional images in holography using inverse imaging, *Optics Express* **16**, 17215–17226 (2008).
29. H. Kim, S.-W. Min, B. Lee, and T.-C. Poon, Optical sectioning for optical scanning holography using phase-space filtering with Wigner distribution functions, *Applied Optics* **47**, D164–D175 (2008).
30. E. Y. Lam, X. Zhang, H. Vo, T.-C. Poon, and G. Indebetouw, Three-dimensional microscopy and sectional image reconstruction using optical scanning holography, *Applied Optics* **48**, H113–H119 (2009).
31. J. Hahn, S. Lim, K. Choi, R. Horisaki, and D. J. Brady, Video-rate compressive holographic microscopic tomography, *Optics Express* **19**, 7289–7298 (2011).
32. F. Zhao, X. Qu, X. Zhang, T.-C. Poon, T. Kim, Y. S. Kim, and J. Liang, Solving inverse problems for optical scanning holography using an adaptively iterative shrinkage-thresholding algorithm, *Optics Express* **20**, 5942–5954 (2012).
33. K. Itoh, Analysis of the phase unwrapping algorithm, *Applied Optics* **21**, 2470–2470 (1982).
34. D. C. Ghiglia, and M. D. Pritt, *Two-Dimensional Phase Unwrapping: Theory, Algorithms, and Software* (John Wiley & Sons, New York, 1998).
35. T. R. Judge, and P. J. Bryanston-Cross, A review of phase unwrapping techniques in fringe analysis, *Optics and Lasers in Engineering* **21**, 199–239 (1994).
36. I. Yamaguchi, T. Ida, M. Yokota, and K. Yamashita, Surface shape measurement by phase-shifting digital holography with a wavelength shift, *Applied Optics* **45**, 7610–7616 (2006).
37. D. Parshall, and M. K. Kim, Digital holographic microscopy with dual-wavelength phase unwrapping, *Applied Optics* **45**, 451–459 (2006).
38. G. Pedrini, P. Fröning, H. J. Tiziani, and M. E. Gusev, Pulsed digital holography for high-speed contouring that uses a two-wavelength method, *Applied Optics* **38**, 3460–3467 (1999).
39. C. Wagner, W. Osten, and S. Seebacher, Direct shape measurement by digital wavefront reconstruction and multiwavelength contouring, *Optical Engineering* **39**, 79–85 (2000).
40. A. Wada, M. Kato, and Y. Ishii, Multiple-wavelength digital holographic interferometry using tunable laser diodes, *Applied Optics* **47**, 2053–2060 (2008).
41. A. Wada, M. Kato, and Y. Ishii, Large step-height measurements using multiple-wavelength holographic interferometry with tunable laser diodes, *Journal of the Optical Society of America A* **25**, 3013–3020 (2008).

42. I. Yamaguchi, S. Ohta, and J.-I. Kato, Surface contouring by phase-shifting digital holography, *Optics and Lasers in Engineering* **36**, 417–428 (2001).
43. D. Velásquez Prieto, and J. Garcia-Sucerquia, Three-dimensional surface contouring of macroscopic objects by means of phase-difference images, *Applied Optics* **45**, 6381–6387 (2006).
44. I. Yamaguchi, J.-I. Kato, and H. Matsuzaki, Measurement of surface shape and deformation by phase-shifting image digital holography, *Optical Engineering* **42**, 1267–1271 (2003).
45. S. Seebacher, T. Baumbach, W. Osten, and W. P. O. Jueptner, Combined 3D-shape and deformation analysis of small objects using coherent optical techniques on the basis of digital holography, *SPIE Proceedings* **4101**, 510–521 (2000).
46. S. M. Solís, F. M. Santoyo, and M. D. S. Hernández-Montes, 3D displacement measurements of the tympanic membrane with digital holographic interferometry, *Optics Express* **20**, 5613–5621 (2012).

7

Computer-generated holography

Computer-generated holography deals with the methods used for the generation of holograms digitally. The hologram can be subsequently printed on a film or loaded onto a spatial light modulator (SLM) for holographic reconstruction. *Computer-generated holograms* (CGHs) have the advantage that the three-dimensional objects do not have to exist in the real world. In other words, the objects which one wants to display can be fictitious. For generating CGHs, different calculation methods have been developed for fitting various display devices and reconstruction methods. In this chapter, we first discuss some of the historical methods of generating computer-generated holograms and then we describe some modern approaches for fast calculations as well as ways to process holographic information. Finally, we discuss three-dimensional holographic display and address some of the issues involving display with SLMs.

7.1 The detour-phase hologram

In the early history of computer-generated holography, neither gray-tone plotters nor SLMs were available. Therefore, it was necessary to reconstruct a quality image from a binary, amplitude-only hologram. The first such 2D CGH was the Fourier-type detour-phase hologram, which was invented by Brown and Lohmann [1, 2].

Figure 7.1 shows the setup for reconstructing a Fourier hologram through a Fourier transform lens of focal length f. We are not concerned with the zeroth-order light and the twin image in the following discussion. As we know, the hologram $H(x, y)$ and the complex amplitude of the reconstructed image $\psi(x, y)$ are related by

$$\psi(x,y) = \mathcal{F}\{H(x,y)\,\psi_r(x,y)\}_{k_x=\frac{k_0 x}{f},\ k_y=\frac{k_0 y}{f}}, \tag{7.1}$$

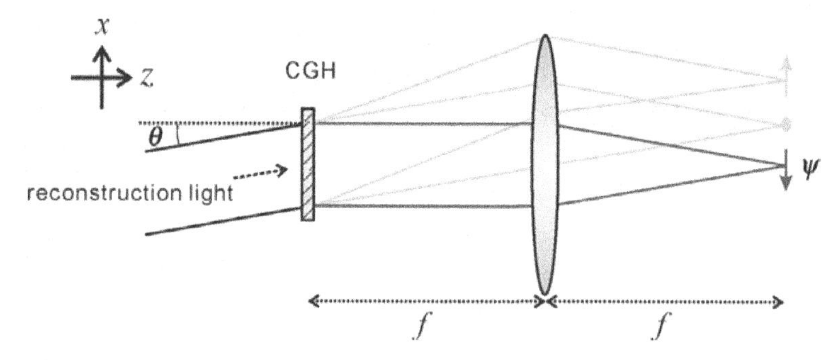

Figure 7.1 Reconstruction scheme for the detour-phase hologram.

where $\psi_r(x, y) = \exp(-jk_0 \sin\theta x)$ denotes the complex amplitude of a plane wave reconstruction light, and θ is the tilt angle.

In general, the hologram is a complex function, that is $H(x, y) = a(x, y) \exp[-j\phi(x, y)]$, where $a(x, y)$ and $\phi(x, y)$ are the modulus and the phase of the hologram, respectively. Hence the goal is to devise a binary (opaque or transparent) amplitude pattern $H_b(x, y)$ to approximate the complex hologram $H(x, y)$. This seems a formidable problem, but if our concern is limited to a window of size $d_x \times d_y$ at the reconstruction plane, it is possible to find $H_b(x, y)$ such that

$$\mathcal{F}\{H_b(x,y)\psi_r(x,y)\} \approx \mathcal{F}\{H(x,y)\psi_r(x,y)\} \tag{7.2}$$

within the window.

First, the area of the hologram is divided into an array of unit cells, and the size of each cell is $w \times w$, as shown in Fig. 7.2. For the complex hologram $H(x, y)$, suppose that both $a(x, y)$ and $\phi(x, y)$ vary slowly within any cell; the cells can then be replaced by an array of point sources, yielding a sampled complex hologram as

$$H_s(x,y) = \sum_{m,n} a_{mn} e^{-j\phi_{mn}} \times \delta(x - x_m, y - y_n), \tag{7.3}$$

where (m, n) indexes the cell centered at $x_m = mw$, $y_n = nw$; $a_{mn} = a(x_m, y_n)$ and $\phi_{mn} = \phi(x_m, y_n)$. Under the illumination of a tilted plane wave, each point source in the sampled hologram produces a plane wave on the reconstruction plane, giving a complex field in the reconstruction plane as

$$\psi_s(x,y) = \mathcal{F}\{\exp(-jk_0 \sin\theta x) H_s(x,y)\}_{k_x = \frac{k_0 x}{f},\ k_y = \frac{k_0 y}{f}} = \sum_{m,n} a_{mn} e^{-j\phi_{mn}} e^{-jk_0 \sin\theta x_m} e^{\left[j\frac{k_0}{f}(x_m x + y_n y) \right]}. \tag{7.4}$$

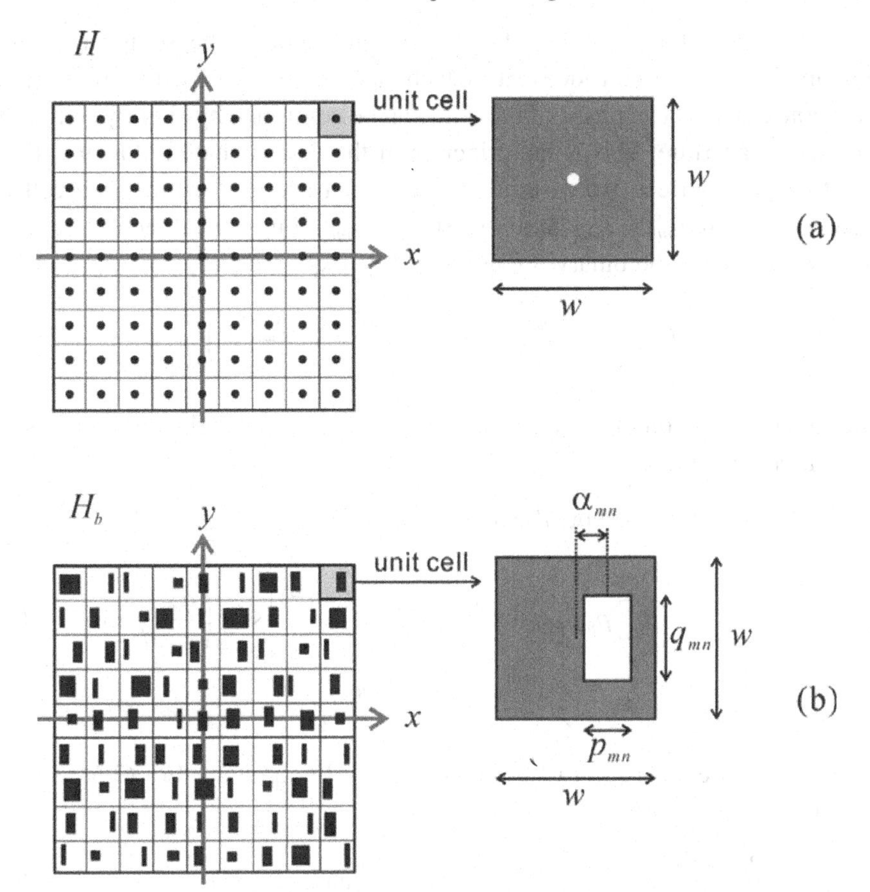

Figure 7.2 Cell description of (a) a complex hologram, and (b) a detour-phase binary hologram.

We can eliminate the effect of the unwanted phase term $\exp(-jk_0 \sin\theta x_m)$ in Eq. (7.4) by designing the width of the unit cell to be

$$w = \frac{2\pi\ell}{k_0 \sin\theta}, \quad \ell \in 1, 2, 3 \ldots \tag{7.5}$$

as $\exp(-j2\pi\ell m) = 1$, yielding

$$\psi_s(x, y) = \sum_{m,n} a_{mn} e^{-j\phi_{mn}} e^{\left[j\frac{k_0}{f}(x_m x + y_n y)\right]}, \tag{7.6}$$

which represents the approximation of a Fourier transform of the sampled complex hologram $H_s(x, y)$.

To understand how a binary hologram can simulate the sampled complex hologram $H_s(x, y)$, we consider that in each cell we open an aperture with an area proportional to a_{mn}. The phase of the light can be controlled by slightly shifting the center of the aperture. This is the principle of the detour-phase hologram. To see the detour-phase effect, we assume that the open aperture within each cell is a rectangle with size $p_{mn} \times q_{mn}$, and is shifted by α_{mn} along the x-direction, as shown in Fig. 7.2(b). Thus the binary hologram can be expressed as

$$H_b(x,y) = \sum_{m,n} \operatorname{rect}\left(\frac{x - x_m - \alpha_{mn}}{p_{mn}}, \frac{y - y_n}{q_{mn}}\right). \tag{7.7}$$

Under the illumination of a tilted plane wave, the complex field at the reconstruction plane is given by

$$\begin{aligned}\psi_b(x,y) &= \mathcal{F}\{\exp(-jk_0 \sin\theta x) H_b(x,y)\}\Big|_{k_x = \frac{k_0 x}{f},\, k_y = \frac{k_0 y}{f}} \\ &= \sum_{m,n} p_{mn} q_{mn} \operatorname{sinc}\left[\frac{p_{mn} k_0 (x - f\sin\theta)}{2\pi f}\right] \operatorname{sinc}\left(\frac{q_{mn} k_0 y}{2\pi f}\right) \\ &\quad \times e^{\frac{jk_0}{f}(x_m x + y_n y)} e^{\frac{jk_0}{f}\alpha_{mn} x} e^{-jk_0 \sin\theta \alpha_{mn}} e^{-jk_0 \sin\theta x_m},\end{aligned} \tag{7.8}$$

where the term, $\exp(-jk_0 \sin\theta x_m)$, becomes unity by applying Eq. (7.5). Furthermore, we assume

$$\frac{p_{mn} k_0 d_x}{2\pi f} \ll 1, \qquad \frac{q_{mn} k_0 d_y}{2\pi f} \ll 1, \quad \text{and} \quad p_{mn} \sin\theta \ll \lambda_0, \tag{7.9}$$

so that the two sinc functions in Eq. (7.8) approach unity within the region of the observation window, where the size of the reconstruction plane is $d_x \times d_y$. Additionally, the term $\exp[j(k_0/f)\,\alpha_{mn} x] \approx 1$ can also be dropped, provided $(k_0/f)\alpha_{mn} d_x \ll 1$. Then, Eq. (7.8) can be simplified to

$$\psi_b(x,y) \approx \sum_{m,n} p_{mn} q_{mn} e^{-jk_0 \sin\theta \alpha_{mn}} e^{\left[j\frac{k_0}{f}(x_m x + y_n y)\right]}. \tag{7.10}$$

By comparing Eq. (7.6) with Eq. (7.10), we can set

$$p_{mn} q_{mn} = c a_{mn}, \tag{7.11a}$$

and

$$k_0 \sin\theta \alpha_{mn} = \phi_{mn}, \tag{7.11b}$$

where c is a proportionality constant. In this way, in the observation window, the binary hologram can reconstruct an image as a complex hologram. In the above

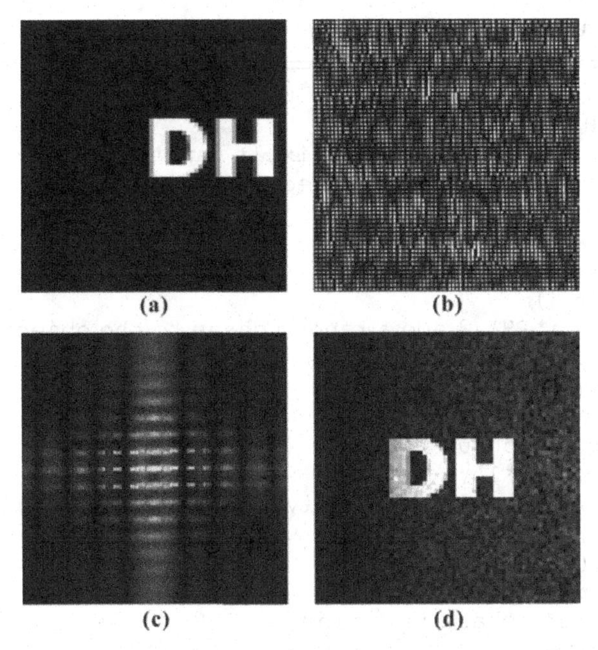

Figure 7.3 (a) Original object, (b) detour-phase hologram of (a), (c) reconstructed image, and (d) enlarged reconstructed image with selected diffracted order from (c).

analysis, the aperture in the cell is rectangular, but it can be any shape, such as circular, with an area proportional to a_{mn}. From Eqs. (7.5) and (7.11b), we can determine the shift of the aperture within each cell by

$$\alpha_{mn} = \frac{\phi_{mn}}{k_0 \sin\theta} = \frac{\phi_{mn}}{2\pi}\frac{w}{\ell}. \tag{7.12}$$

Because ϕ_{mn} is wrapped between π and $-\pi$, the range of α_{mn} is within $\pm w/2\ell$.

Example 7.1: Generation of a detour-phase CGH

In this example we will generate a detour-phase CGH. In the detour-phase hologram, the structure in a cell must be controlled. So we use 16 × 16 pixels to construct a cell, and the hologram consists of 64 × 64 cells. So there is a total of 1024 × 1024 pixels in the hologram. The 64 × 64-cell object is shown in Fig. 7.3(a). It is noted that we must attach a random-phase mask against the object in order to broaden its spectrum. Otherwise, most of the energy will concentrate in the zeroth-order beam and the effective dynamic range of the produced hologram will be reduced. The random-phase-added object is transformed to the Fourier domain also with 64 × 64 cells. The value of each cell is complex, and will

Table 7.1 *MATLAB code for generating a detour-phase CGH, see Example 7.1*

```
clear all; close all;
I=imread('DH64.bmp', 'bmp');
% please prepare a 8 bit, 64x64 pixels input image
% otherwise the code must adjusted accordingly
I=double(I);
figure; imshow(abs(I));
title('Original object')
PH=rand([64,64]);
I=I.*exp(2i*pi*PH);% add a random phase to the object
FTS=fftshift(ifft2(fftshift(I)));
A=abs(FTS);
figure;imshow(mat2gray(A));
title('Object spectrum')
A=A./max(max(A))*15;
A=round(A);% The amplitude is divided into 16 levels
B=angle(conj(FTS));
B=B-min(min(B));
B=B./max(max(B))*7;
B=round(B);% The phase is divided into 8 levels
H=zeros(1024);
 for m=1:64;
     for n=1:64;
         P=zeros(16);
         a=A(m,n);
         b=B(m,n);
         c=fix(a/2);
         d=rem(a,2);
         P(9-c:8+c+d,(1+b):(9+b))=1;
         H(16*(m-1)+1:16*(m-1)+16,16*(n-1)+1:16*(n-1)+16)=P;
     end
 end
figure; imshow(H)
title('Detour-phase CGH')
imwrite(H, '1AA.jpg','jpg')

%Reconstruction (FFT)
R=fftshift(ifft2(fftshift(H)));
figure; imshow(100.*mat2gray(abs(R)));
title('Reconstructed image')
```

be coded to a binary pattern as a cell. The MATLAB code is listed in Table 7.1 as a reference.

First, the modulus and the phase are quantized to 16 and 8 levels, respectively. We set $p_{mn} = 9$ pixels, and $q_{mn} = 1 \sim 16$ pixels to represent 16 modulus levels. We also select $\alpha_{mn} = -4 \sim 3$ pixels to represent 8 phase levels. In other words, we select $\ell = 2$ in Eqs. (7.5) and (7.12). By using this setup, the opening of the aperture will always be inside a cell. The produced detour-phase hologram is shown in Fig. 7.3(b). The reconstructed image is shown in Fig. 7.3(c), while a selected diffracted order is enlarged in Fig. 7.3(d). There are many diffracted orders of light due to the cell grids, but only specific diffracted orders satisfy the assumptions in our analysis, resulting in quality images. Note that in the example the diffraction efficiency is very low because we use a simple but low-efficiency coding pattern. For example, p_{mn} cannot exceed 9 pixels while the cell width, w, is 16 pixels in our present example. One may devise other coding patterns to improve the diffraction efficiency.

7.2 The kinoform hologram

Although the detour-phase hologram can represent the phase as well as the modulus of a sampled complex hologram, its diffraction efficiency is relatively low. To achieve high diffraction efficiency, the kinoform hologram has been proposed [3]. The kinoform hologram is also based on the scheme of Fourier holography. Generally speaking, the Fourier hologram $H(x,y) = a(x,y)\exp[-j\phi(x,y)]$ is a complex hologram. We mentioned in Example 7.1 that a random-phase mask is usually attached against the object pattern to spread the energy across a wide spectrum. If the energy spread is uniform, the modulus $a(x, y)$ is relatively unimportant and can be ignored. So we only need to extract the phase $\phi(x, y)$ and produce a gray-tone pattern whose gray-level is proportional to $\phi(x, y)$. The gray-tone pattern can be displayed on a phase-only spatial light modulator or printed on a photographic film. If a gray-tone pattern is to be produced on a photographic film, the film is bleached and the bleaching is a chemical process for converting an amplitude hologram into a phase hologram. In this way, we can generate a kinoform hologram given by

$$H_p(x, y) = \exp[-j\phi(x, y)]. \tag{7.13}$$

The design of a kinoform hologram is very simple. Ideally, there is only one diffraction order emitting from the kinoform hologram. Thus the kinoform hologram can be an on-axis hologram. Moreover, the diffraction efficiency of the kinoform hologram can approach 100%. However, the fabrication of the kinoform hologram is not as simple as its design. First, the gray-tone fringes of the amplitude

Figure 7.4 (a) Original object, (b) random-phase mask with phase ranging from $-\pi$ to π, (c) spectrum of the object with the random-phase mask, and (d) reconstructed image of the kinoform (phase-only) hologram.

hologram must be plotted correctly. Second, the amplitude transmittance must be converted to phase retardation linearly. And finally, the phase retardation must be precise in the range $0 \sim 2\pi$ (integral multiples of 2π radians are subtracted for any point in the hologram to be within 2π radians). Otherwise, the diffraction efficiency drops and the zeroth-order light appears. In computer-generated holography, kinoform holograms are easy to generate as we simply extract the phase information from the computed complex field of an object and display on a phase-only SLM. To demonstrate the principle, we use a pattern shown in Fig. 7.4(a) as the input. We also attach a random-phase mask against the pattern. The random-phase mask is shown in Fig. 7.4(b). The phase of the spectrum of the object with the random-phase mask is shown in Fig. 7.4(c). Although the energy spreads out by the random phase in the spectrum domain, the spectrum is locally non-uniform. While we still assume a uniform modulus to generate a kinoform hologram, the corresponding reconstructed image is shown in Fig. 7.4(d). The MATLAB code is listed in Table 7.2. There is notable speckle noise in the reconstructed image. This is a consequence of the non-uniform modulus of the spectrum. The problem can be avoided by using specific object patterns. Alternatively, optimization procedures, such as the iterative Fourier transform algorithm (Section 7.3), can be applied to improve the image quality [4, 5].

Table 7.2 *MATLAB code for generating a kinoform computer-generated hologram, see Fig. 7.4*

```
clear all; close all;
I=imread('cameraman.tif', 'tif');
I=double(I);
figure; imshow(mat2gray(I));
title('Original object')
PH=rand([ 256,256] );
I=I.*exp(2i*pi*PH);% add a random phase to the object
FTS=fftshift(ifft2(fftshift(I)));
A=abs(FTS);
figure; imshow(mat2gray(A));
title('Spectrum modulus')
B=angle(FTS);
figure; imshow(mat2gray(B));
title('Spectrum phase')

%Reconstruction (FFT)
R=fftshift(ifft2(fftshift(exp(-1j*B))));
figure; imshow(mat2gray(abs(R)));
title('Reconstructed image')
```

7.3 Iterative Fourier transform algorithm

In Section 7.2, we introduced the kinoform hologram. The diffraction efficiency of the kinoform hologram is greater than that of the detour-phase hologram because kinoform holograms are phase only. However, the resulting image reconstructed from the kinoform hologram is usually noisy. In this section, we introduce the *iterative Fourier transform algorithm* (IFTA), which is able to optimize computer-generated phase-only Fourier holograms. The IFTA was first proposed by Hirsch *et al.* [6]. Independently, Gerchberg and Saxton dealt with the phase-retrieval problem using a similar algorithm [7]. So the IFTA is also called the *Gerchberg–Saxton algorithm.*

In generating a phase-only hologram, $H_p(x, y)$, the amplitude of the reconstructed image must be proportional to the desired amplitude distribution $A(x, y)$, that is

$$\mathcal{F}\{H_p(x,y)\} = \mathcal{F}\{e^{-j\phi(x,y)}\} = A(x,y)e^{-j\theta(x,y)}, \tag{7.14}$$

where $\phi(x,y)$ is the phase of the hologram, and $\theta(x,y)$ is the phase of the reconstructed light. In designing the phase-only CGH, we need to find a phase function $\theta(x, y)$ that will let H_p become a phase-only function. Based on this idea, the IFTA is an algorithm for searching for the best solution of $\theta(x,y)$.

Figure 7.5 shows a general flowchart of the IFTA for generating a Fourier hologram. First, an initial pattern is loaded as the amplitude (modulus) distribution

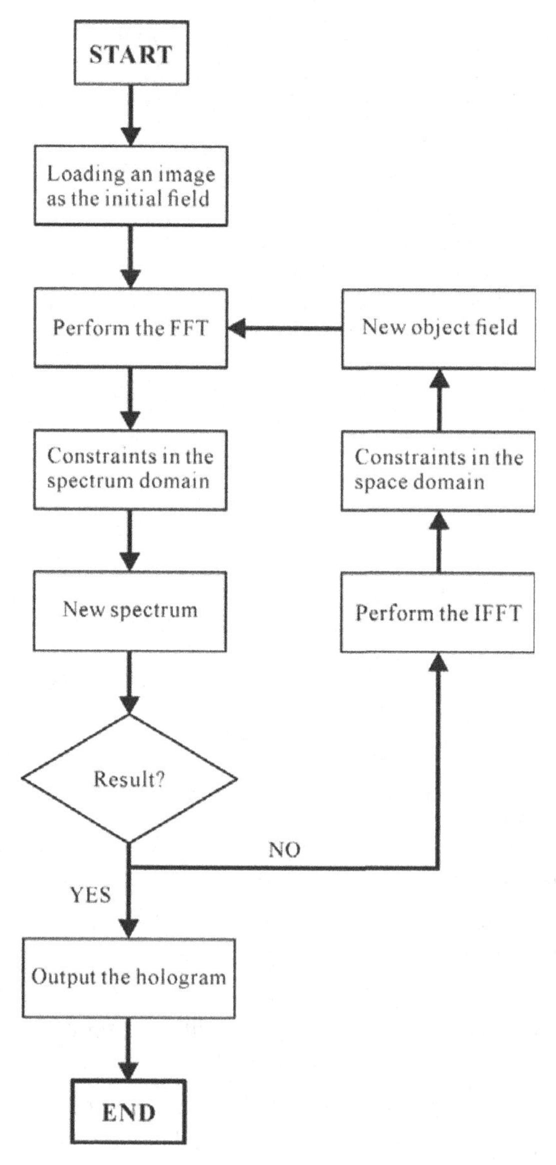

Figure 7.5 Flowchart of the iterative Fourier transform algorithm (IFTA).

$A(x, y)$ of the target reconstructed field. If desired, one can also apply a phase mask on the input pattern as the initial field. The initial field is then Fourier transformed to the spectrum domain. The spectrum is modified according to a constraint. In our problem the constraint is that the spectrum modulus must be uniform. So we set the spectrum modulus to become unity without altering its phase. The modified spectrum is transformed back to the spatial domain. We then apply a constraint on the resulting field in the spatial domain, thereby obtaining a new field. In our problem the constraint in the spatial domain is the given modulus distribution, $A(x, y)$. So we enforce the modulus of the field to be $A(x, y)$ without altering its phase. The resulting field is then regarded as the initial field in the next iteration. Iterations are repeated until the goal (e.g., iteration number) is achieved.

When using the IFTA, it is important to hold sufficient degrees of freedom. If the constraint in either the spatial domain or the spectrum domain is too severe, there will be no satisfactory solution. If the constraint is proper, the resulting field will converge to an optimized solution. Note that the IFTA only searches for a small region of the phase-function set. For example, for a 512×512-pixel phase function with 256 phase levels, there are a total of $256^{512\times512}$ phase functions. Thus, the solution found by the IFTA may be only locally optimized, i.e., it is the best solution within the searched region, but not a globally optimized solution (the best solution in the complete set). Additional algorithms must be included in the IFTA to search for a globally optimized solution [8, 9].

Example 7.2: Phase-only Fourier hologram by IFTA

In this example, we use the image shown in Fig. 7.4(a) as the initial pattern. The MATLAB code for the IFTA is listed in Table 7.3. First, the initial input image is transformed to the spectrum domain, and the phase spectrum is extracted and transformed back to the spatial domain. To monitor the quality of the resulting reconstructed image of the hologram, we measure the *root-mean-square error (RMSE)*, which is defined as

$$\text{RMSE} = \left\{ \frac{1}{MN} \sum_{m,n} \left[|\psi(m,n)| - A(m,n) \right]^2 \right\}^{0.5}, \tag{7.15}$$

where $A(m, n)$ is the target image, $\psi(m, n)$ is the evaluated field, (m, n) are the sampling indices, and M and N are the sampling numbers along the x-axis and y-axis, respectively. Figure 7.6(a) shows the RMSE as a function of the iteration number. In the example, the total number of iterations is 100. The error decreases gradually as the number of iterations increases. The resulting reconstructed image after 100 iterations is shown in Fig. 7.6(b). Note that it is comparable to Fig. 7.4(a).

7.4 Modern approach for fast calculations and holographic information processing

7.4.1 Modern approach for fast calculations

Holography is a technique for recording the wavefront of a three-dimensional scene on a two-dimensional recording device such as a photographic film or CCD camera. A few decades ago, the recording process could only be accomplished by optical means by mixing the object wave with a reference wave, and recording the resulting

Table 7.3 *MATLAB code for generating a phase-only Fourier hologram using the IFTA, see Example 7.2*

```
clear all; close all;
I=imread('cameraman.tif', 'tif');
I=double(I);
I=I./max(max(I));
avg1=mean(mean(I));
figure;imshow(mat2gray(I));
title('Original object');
figure;
axis([0,101,0,1]);
xlabel('Number of iterations')
ylabel('RMSE')
hold on
I1=I;
for n=1:100; % iterations to optimize the phase hologram
H=fftshift(fft2(fftshift(I1)));
I2=fftshift(ifft2(fftshift(exp(1j.*angle(H)))));
avg2=mean(mean(abs(I2)));
I2=(I2./avg2).*avg1;
rmse=(mean(mean((abs(I2)-I).^2)))^0.5;
plot(n,rmse,'o');
pause(0.3); % To see the error in each iteration.
I1=I.*exp(1j*angle(I2));
end
hold off
I2=I2./max(max(abs(I2)));
figure;imshow(abs(I2));
title('Reconstructed image')
```

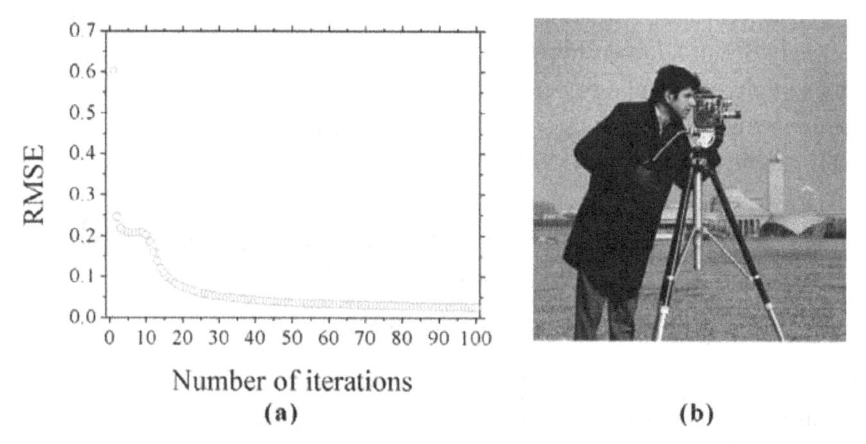

Figure 7.6 (a) RMSE as a function of iteration number, (b) reconstructed image after 100 iterations.

fringe pattern on a recording device. With the rapid advancement of computing technologies in recent years, the optical hologram formation process can now be simulated by numerical means. Such an approach, commonly referred to as *computer-generated holography* [1], computes the diffraction pattern emitted from a three-dimensional object and adds a reference wave to produce a digital hologram. The hologram can be displayed on electronic addressable devices (e.g., spatial light modulator) [10], or printed with laser [11] or fringe printers [12], and subsequently reconstructed to give a three-dimensional scene upon suitable illumination. Digital holography is likely to be a promising solution in the next generation of three-dimensional display. At the same time, such optimism is masked by a number of practical problems which are difficult to solve in the foreseeable future. Many of these problems can be traced back to the fine pixel size of a hologram, which is of the order of the wavelength of light. For example, a small 10 mm × 10 mm hologram with a square pixel size of 5 μm × 5 μm, is made up of over 4×10^6 points, which is about two times more than the number of pixels in a domestic high-definition television of 1920 × 1080 pixels. One can easily imagine the formidable amount of computation associated with the generation and processing of a large hologram digitally. In this section, we will first discuss research based on a modern framework of digital holography known as the wavefront recording plane (WRP), which is generalized as the virtual diffraction plane (VDP), if digital processing of holographic data is desired. We will then cover the framework based on the VDP in the subsequent section.

The concept and theory of the WRP were first proposed by a group of researchers into fast generation of digital holograms [13, 14]. Traditional approaches (for example, [15–17]) aimed at enhancing the speed of generating a hologram directly from an object scene, whereas the WRP approach considers a hypothetical virtual window that is parallel to the plane of the hologram and placed at close proximity to the object scene [18]. The situation is shown in Fig. 7.7.

An effective method in computer-generated holography is the point-light concept, which is a numerical realization of the zone-plate approach to holography

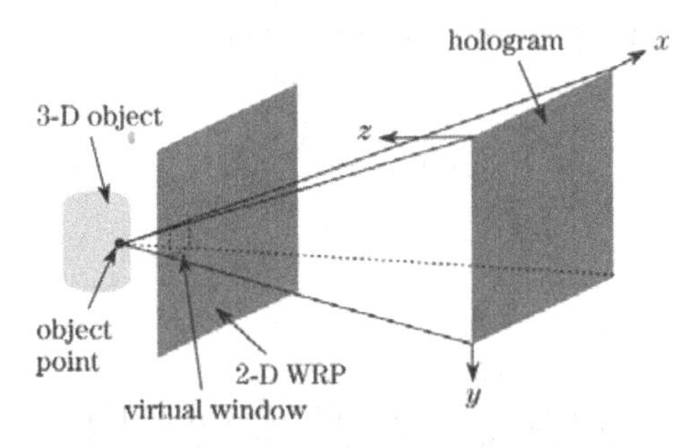

Figure 7.7 Spatial relation between the 3-D object space, the 2-D WRP, and the hologram. From Ref. [18], with permission, © *Chinese Optics Letters*.

[19]. Briefly, an object is assumed to be composed of self-illuminated points each contributing to an elementary hologram known as a Fresnel zone plate (FZP). The hologram of a three-dimensional object is generated by summing the FZP corresponding to each individual object point. A hologram generated using this method is known as a Fresnel hologram. Given a set of three-dimensional object points

$$O(x,y,z) = \left[o_0\left(x_0,y_0,z_0\right), o_1\left(x_1,y_1,z_1\right), \ \ldots, \ o_{N-1}\left(x_{N-1},y_{N-1},z_{N-1}\right)\right].$$

The intensity of each point, and its axial distance (depth) from the hologram are given by a_i and z_i, respectively. The Fresnel hologram, $H(x, y)$, is given by

$$H(x,y)\Big|_{\substack{0 \le x < X \\ 0 \le y < Y}} = \sum_{i=0}^{N-1} \frac{a_i}{r_i} \exp\left(-j\frac{2\pi}{\lambda_0} r_i\right), \tag{7.16}$$

where X and Y are the horizontal and vertical extents of the hologram, respectively, and are assumed to be identical to that of the object scene. λ_0 is the wavelength of the optical beam which is used to generate the complex hologram. The term $r_i = \sqrt{(x-x_i)^2 + (y-y_i)^2 + z_i^2}$ is the distance of an ith object point at position $(x_i, y_i; z_i)$ to a point at (x, y) on the hologram. From Eq. (7.16), we can see that each object point is contributing to the entire hologram, and the evaluation of each hologram pixel involves the complicated expression enclosed in the summation operation. In the framework of the WRP, as evident from Fig. 7.7, wavefronts emitted from a self-illuminating point will diverge to cover the entire hologram and intercept the WRP in its path. If the WRP is near to the object point, the coverage of the object wavefronts on the WRP is limited to a small virtual window. For simplicity, the virtual window is assumed to be a square of size $W \times W$. The Fresnel diffraction equation in Eq. (7.16) can be applied, with slight modification, to compute the diffraction pattern $u_w(x, y)$ on the WRP within the virtual window. Let d_i denote the axial distance from the ith object point to the WRP, then we have the field distribution on the virtual window given by

$$u_w(x,y) = \sum_{i=0}^{N-1} F_i \tag{7.17}$$

where

$$F_i = \begin{cases} \dfrac{a_i}{R_{wi}} \exp\left(-j\dfrac{2\pi}{\lambda_0} R_{wi}\right) & \text{for } |x-x_i| \text{ and } |y-y_i| < \frac{1}{2}W \\ 0 & \text{otherwise,} \end{cases}$$

and where $R_{wi} = \sqrt{(x-x_i)^2 + (y-y_i)^2 + d_i^2}$ is the distance of the point from the WRP to the ith object point. In Eq. (7.17), computation of the WRP for each object point is only confined to the region of the virtual window on the WRP. As W is much smaller than X and Y, the computation load is significantly reduced compared with that in Eq. (7.16). In Ref. [15], the calculation is further simplified by pre-computing the terms $(1/R_{wi}) \exp(-j(2\pi/\lambda_0)R_{wi})$ for all combinations of $|x - x_i|$, $|y - y_i|$, and d_i (within the coverage of the virtual window), and storing the results in a look-up table (LUT). In the second stage, the field distribution in the WRP is expanded to the hologram plane given by

$$u(x, y) = \mathcal{F}^{-1}\{\mathcal{F}\{u_w(x, y)\}\mathcal{F}\{h(x, y; z_w)\}\}, \tag{7.18}$$

where z_w is fixed for a given separation between the WRP and the hologram. As demonstrated by Shimobaba *et al.* [14], a video sequence of digital holograms, each comprising 2048×2048 pixels and representing 10^{14} object points, can be generated at a rate of 10 frames per second.

Adopting the WRP framework, Tsang *et al.* proposed a method known as the interpolated wavefront recording plane (IWRP) method [20], and extended it to handle not just a plane object surface but also a three-dimensional object surface which has the same size and number of pixels as the hologram. In this approach, it is assumed that the resolution of the object scene is generally smaller than that of the hologram. Hence it is unnecessary to convert every object point of the scene to its wavefront on the WRP. On this basis, the object scene is evenly partitioned into a non-overlapping square support of size $M \times M$ as shown in Fig. 7.8. The object scene is sampled at the sample point (m, n), which is at the center of the square support.

This is equivalent to subsampling the object scene evenly M times along the horizontal and the vertical directions. It is then assumed that each sample point is contributing to the wavefront only to a square virtual window in the WRP as shown in Fig. 7.9. Although the reduction has effectively reduced the computation time, as shown in Ref. [20], the reconstructed images obtained are weak, noisy, and difficult to observe upon optical reconstruction. This is of course caused by the sparse distribution of the object points caused by subsampling of the scene image. To overcome this issue, the interpolated WRP (IWRP) is proposed to interpolate the support of each object point with padding, i.e., the object point is duplicated to all the pixels within each square support. After the interpolation, the wavefront of a virtual window will be contributed by all the object points within the support. Note that since each object point is duplicated to all the pixels within each square support, the wavefront on the visual window from each interpolated object point is simply a shifted version of the wavefront from the sample point, which can be pre-computed and stored in a look-up table (LUT). Consequently, each virtual window

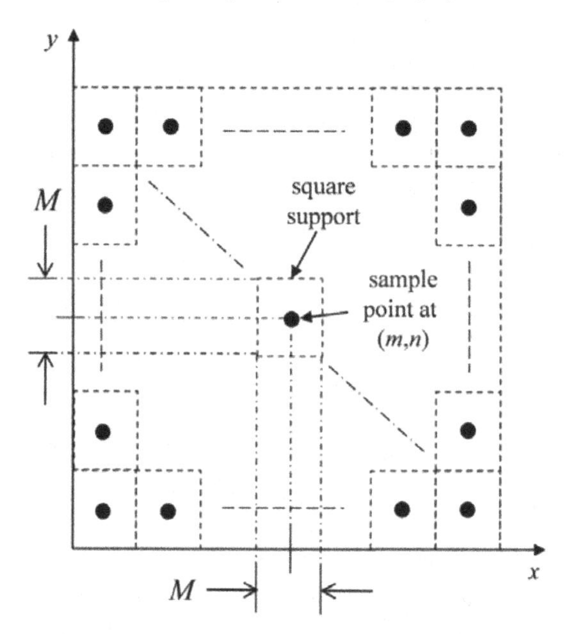

Figure 7.8 Partitioning of the object scene into a non-overlapping square support.

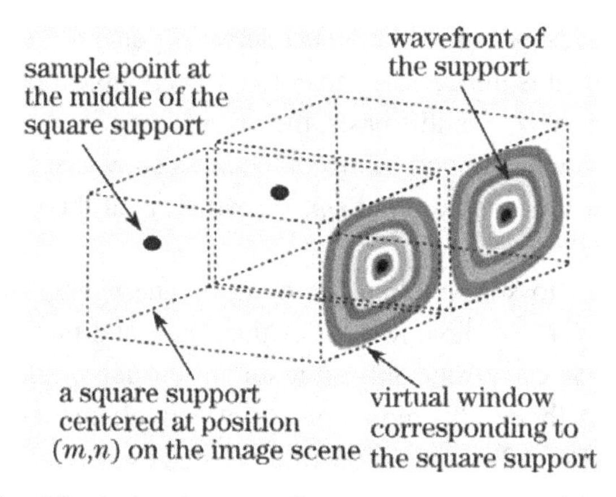

Figure 7.9 Spatial relation between the square support and its corresponding virtual window. From Ref. [18], with permission, © *Chinese Optics Letters.*

in the IWRP can be generated in a computation-free manner by retrieving from the LUT the wavefront corresponding to the intensity and depth of the corresponding object point. The method is capable of generating a hologram representing an object of over 10^6 points over 40 frames per second – a state-of-the-art experimental result in CGH [20].

Figure 7.10 shows some experimental results using the WRP and IWRP methods. Figure 7.10(a) is an image divided (dotted line) into a left and a right

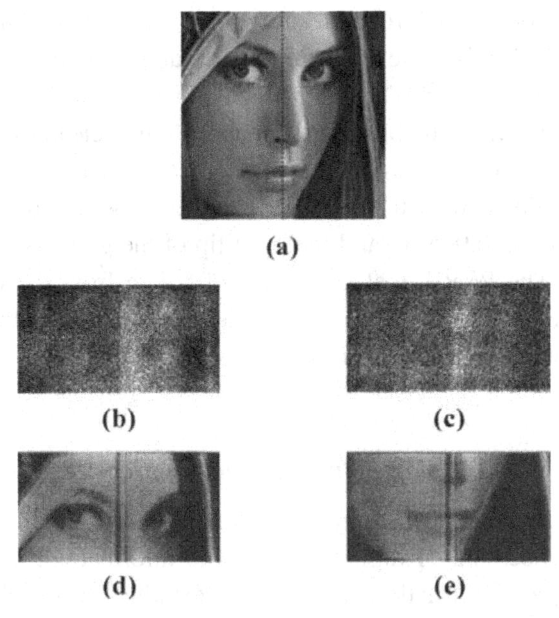

Figure 7.10 (a) An image divided (dotted line) into left and right parts, positioned at $z_1 = 0.005$ m (left) and $z_2 = 0.01$ m (right) from the WRP/IWRP, respectively. The WRP/IWRP is positioned at $z_w = 0.4$ m from the hologram; (b) optically reconstructed upper half of the image using the WRP method, (c) same as (b) but for the lower half of the image, (d) optically reconstructed upper half of the image using the IWRP method, and (e) same as (d) but for the lower half of the image. From Ref. [20], with permission, © OSA.

part, positioned at distances of $z_1 = 0.005\,\text{m}$ and $z_2 = 0.01\,\text{m}$ from the WRP/IWRP. The image is of 2048 × 2048 square pixels each with a size of $9 \times 9\ \mu\text{m}^2$ and quantized with 8 bits. Each pixel in the image is taken to generate the hologram, constituting a total of around 4×10^6 object points. Figure 7.10 (b) to (d) are optically reconstructed real off-axis holograms. The hologram $\mathcal{H}(x, y)$ is generated by multiplying a planar reference wave $e^{-jk_0 y \sin\theta}$ (illuminating at an inclined angle $\theta = 1.2°$ on the hologram) by $u(x, y)$, and taking the real part of the result given by

$$\mathcal{H}(x, y) = \text{Re}\left\{u(x, y)e^{-jk_0 y \sin\theta}\right\}. \tag{7.19}$$

The hologram is displayed on a liquid crystal on silicon (LCOS) SLM modified from the Sony VPL-HW15 Bravia projector. The projector has a horizontal and vertical resolution of 1920 and 1080, respectively. Due to the limited size and resolution of the LCOS SLM, only part of the hologram (and hence the reconstructed image) can be displayed. Employing the WRP method, the reconstructed images corresponding to the upper half and the lower half of the hologram are shown in Figs. 7.10(b) and 7.10(c), respectively. We observe that the images are extremely weak and noisy. Next we repeat the above process using the IWRP method with $M \times M = 8 \times 8$. The reconstructed images are shown in Figs. 7.10(d) and 7.10(e). Evidently, the reconstructed image is much clearer in appearance. To illustrate further the usefulness of the method, a sequence of holograms of a rotating globe rendered with the texture of the Earth was generated. The radius of the globe is around 0.005 m, and the front tip of the globe is located at 0.01 m from the IWRP. The IWRP is at a distance of 0.4 m from the hologram. For a hologram (as well as image size) of 2048×2048 pixels, the method is capable of attaining a generation speed of over 40 frames per second [20].

7.4.2 Holographic information processing

Aside from the rejection of the zeroth-order light and the twin image, in recent years, the post-processing of digital holographic information, i.e., processing of recorded holographic information, has been most studied. Research in this area, for example sectioning in digital holography, seems to have started in 2006 [21–23] and recent work on optimization using the L_2 norm method has led to impressive sectioning results [24], which has motivated the use of the L_1 norm, eventually leading to what is commonly known as *compressive holography* [25]. In this section, we discuss recent innovations in post-processing. Post-processing is based on the principle of the virtual diffraction plane (VDP) – a generalization of the WRP previously discussed. The principle of VDP-based holographic processing has three stages as shown in Fig. 7.11.

The conceptual idea of VDP-based holographic processing can be visualized physically with the help of Fig. 7.7, where we introduced the concept of the WRP. Now, as shown in Fig. 7.11, we first back-project the hologram to a virtual plane called the virtual diffraction plane (VDP). In a sense the WRP is now called the VDP. The VDP is close to the original three-dimensional object. Due to the close proximity, the wave emitted from each object point will only cover a small, localized area on the VDP (in contrast with the wave projected onto the hologram,

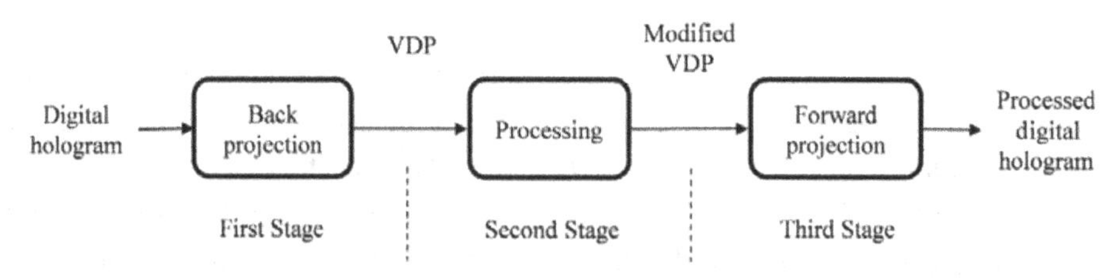

Figure 7.11 Hologram post-processing based on the VDP framework.

which covers the entire area). As a result, the overall diffraction pattern on the VDP, which is the summation of the contribution of individual object points, can be deduced with a very small amount of computation. The rationale is that the VDP is found to carry the diffraction fringes of the entire hologram, but at the same time it carries similar optical properties as the object scene. Generally speaking, this implies that modifying the brightness and contrast on the VDP will lead to almost identical changes in the pictorial contents it represents. Along this line of thinking, a hologram can revert into a VDP. The information on the VDP can be conveniently processed using many existing image processing techniques and tools. Therefore, in the second stage shown in Fig. 7.11, we perform image processing on the VDP. After processing, the VDP can be easily expanded into the ultimate hologram by forward projection, which is further away from the object scene.

Let us now formulate the idea mathematically. Suppose the VDP is located at an axial distance z_v from the hologram, the complex wavefront $u_{VDP}(x, y)$ on the VDP is back-propagated from the hologram $t(x, y)$ given by

$$u_{VDP}(x,y) = t(x,y) * h(x,y;-z_v) = t(x,y) * h^*(x,y;z_v). \tag{7.20}$$

In the second stage, a selected region(s) on the VDP, denoted by S, is processed with a given function $P[\cdot]$. After the modification, the diffraction pattern on the VDP becomes

$$v(x,y) = \begin{cases} P[u_{VDP}(x,y)] & (x,y) \in S \\ u_{VDP}(x,y) & \text{otherwise.} \end{cases} \tag{7.21}$$

In the third stage of the VDP processing framework, the modified or the processed wavefront $v(x, y)$ is forward projected onto a hologram plane by convolving with the spatial impulse response in Fourier optics given by

$$t_P(x,y) = v(x,y) * h(x,y;z_v). \tag{7.22}$$

We can realize Fourier transformation in Eqs. (7.20) and (7.22), and the Fourier transform of $h(x, y; z_v)$ and $h^*(x, y; z_v)$ can be pre-computed in advance. Hence, the processing of a hologram with the VDP framework mainly involves a pair of forward and inverse Fourier transform operations. The rest of the process is negligible with regard to computation time. Based on a commercial graphic processing unit (GPU) to conduct the Fourier transform, a medium size digital hologram comprising 2048×2048 pixels can be processed in less than 10 ms, equivalent to a rate of over 100 frames per second [26].

Figure 7.12 shows an example of VDP-based processing of the hologram. Apart from readjusting the brightness and contrast of the image represented in a hologram, sometimes it is also necessary to enhance the sharpness in certain regions to increase the visibility of the high frequency contents. This process can be achieved

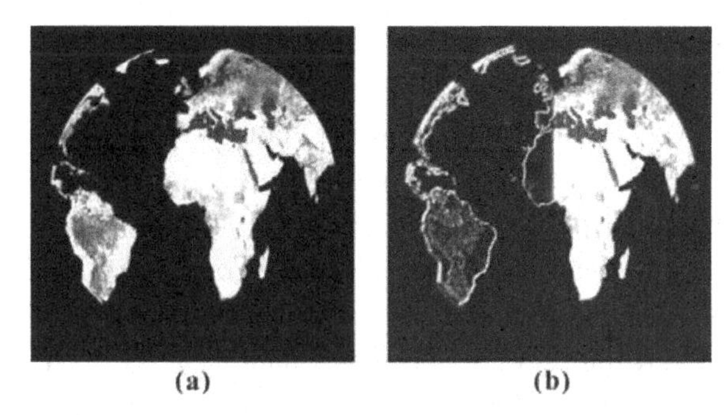

Figure 7.12 (a) A hemisphere rendered with the texture of the Earth positioned at 0.3 m from the hologram plane. (b) Numerical reconstructed image of the hologram representing the image in (a) after direct application of high-boost filtering to the left side of the virtual diffraction plane. From Ref. [26], with permission, © OSA.

by applying a high-boost filter to the selected area on the VDP of the hologram. High-boost filtering is performed by amplifying the original input image and then subtracting a lowpass image, which is given by

$$v(x,y)|_{(x,y)\in S} = \left[Au_{VDP}(x,y) - Bu_{VDP}^{L}(x,y)\right], \tag{7.23}$$

where $u_{VDP}^{L}(x,y)$ is a lowpass version of the region S. Each pixel in $u_{VDP}^{L}(x,y)$ is derived from the average value of a 3×3 window centered at the corresponding pixel according to

$$u_{VDP}^{L}(x,y) = \frac{1}{9}\sum_{m=-1}^{1}\sum_{n=-1}^{1} u_{VDP}(x+m, y+n). \tag{7.24}$$

The terms A and B are constant values. The larger the values of A and B, the higher will be the brightness and sharpness of the region S, respectively. Other types of sharpening filters can be applied under the same principle. The hologram sharpening process is illustrated by a hemisphere with the texture of the Earth as shown in Fig. 7.12. The hemisphere has a radius of 5 mm with its tip located at 0.3 m from the hologram. Once the digital hologram is obtained, the VDP is generated using Eq. (7.20). The high-boost filter in Eq. (7.23) is applied to the VDP, and subsequently forward projected into a hologram using Eq. (7.22). The reconstructed image of the modified hologram, focused at 0.3 m (which causes slight de-focusing around the rim of the hemisphere), is shown in Fig. 7.12(b). It can be seen that the edge on the left side of the reconstructed image is strengthened, and the rest of the reconstructed image is not affected as S, the region to be processed, has been chosen to the left half of the plane.

7.5 Three-dimensional holographic display using spatial light modulators

Static holograms with high fidelity have been achievable for a long time, but the real-time display of high-quality digital holograms remains a challenge due to the limitations of modern spatial light modulators (SLMs). The real-time display of digital holograms demands SLMs of high resolution and with the capability of updating frames at video rate [27]. In this section, we consider some of the issues in real-time holographic display. Of course, other practical issues such as phase uniformity over the SLM surface (which is important as high-quality read-out employs coherent light) and large-area format would also need to be taken into account for any practical real-time display system.

7.5.1 Resolution

To consider the high resolution issue, for simplicity, let us display an on-axis point source hologram as an example. According to Chapter 2, the on-axis point source hologram is given by

$$t(x,y) = A + B\ \sin\left[\frac{k_0}{2z_0}\left(x^2+y^2\right)\right]. \tag{7.25}$$

Again, z_0 is the distance of the point source from the hologram. An instantaneous spatial frequency along the x-direction within the hologram is

$$f_{inst}(x) = \frac{1}{2\pi}\frac{d}{dx}\left[\frac{k_0}{2z_0}x^2\right] = \frac{x}{\lambda_0 z_0}. \tag{7.26}$$

Assuming the SLM is limited to the size x_{max} with spatial resolution f_0, then the highest spatial frequency of the hologram fringes is $f_{inst}\ (x_{max}) = x_{max}/\lambda_0 z_0$ and if we want to record it, we must set

$$f_{inst}(x_{max}) = \frac{x_{max}}{\lambda_0 z_0} = f_0. \tag{7.27}$$

Hence, for a given spatial resolution of the SLM, we can find the limiting aperture, $2x_{max}$, of the hologram on the SLM. We define the numerical aperture of the hologram as

$$NA = \sin\left(\frac{\theta}{2}\right) = \frac{x_{max}}{\sqrt{x_{max}^2 + z_0^2}} = \frac{x_{max}}{z_0}\left[1+\left(\frac{x_{max}}{z_0}\right)^2\right]^{-0.5}, \tag{7.28}$$

where θ is the viewing angle as shown in Fig. 7.13, as the on-axis hologram is reconstructed with a real point source located z_0 away from the hologram.

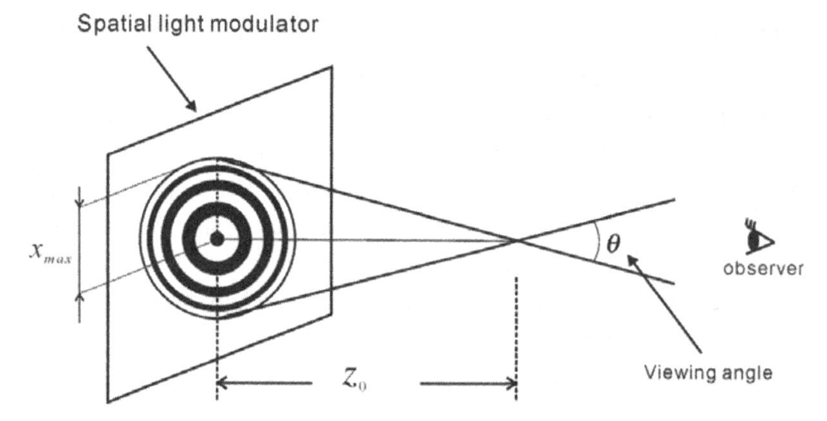

Figure 7.13 Viewing angle for an on-axis point-object hologram.

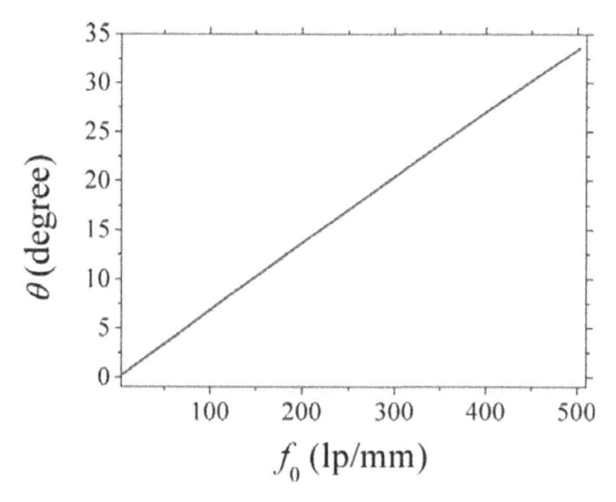

Figure 7.14 Spatial resolution versus viewing angle.

Combining Eqs. (7.27) and (7.28), we can find the viewing angle in terms of the spatial resolution of the SLM as follows:

$$\theta = 2 \times \sin^{-1}\left[\frac{\lambda_0 f_0}{\sqrt{1 + (\lambda_0 f_0)^2}}\right]. \tag{7.29}$$

For an illumination wavelength $\lambda_0 = 0.6$ μm, Fig. 7.14 shows f_0 versus the viewing angle.

For example, Hamamatsu's electron-beam–addressed SLM with a spatial resolution of about $f_0 = 8$ lp/mm (line pairs per millimeter) gives a viewing angle of about 0.6°. Even with Hamamatsu's parallel aligned nematic liquid crystal SLM (PALSLM) with $f_0 = 100$ lp/mm, we have $\theta \approx 6.8°$. In fact, none of the modern

SLMs, typically with a spatial resolution of about 100 lp/mm, are suitable for large viewing angle three-dimensional display. Since 1 lp/mm means there are two pixels in 1 mm, 100 lp/mm means the pixel size is about $\frac{1}{2} \times \frac{1}{100}$ mm or 5 μm for modern SLMs. The situation becomes even worse if off-axis holography is employed because the SLM needs to resolve the carrier frequency. For an offset angle of 45°, the spatial carrier frequency is sin $45°/\lambda_0 \approx 1000$ lp/mm or a pixel size of about 0.5 μm, well beyond modern SLM capability. Hence high resolution holographic display with SLMs is an important area of research.

An effective solution for higher resolution holographic display has been realized through the integration of optically addressed SLMs and the active tilting method [28, 29]. But both the cost and complexity of such systems are high. Here we discuss a modern approach to address the issue of low resolution SLMs. The type of novel digital hologram generated is called a *binary mask programmable hologram (BMPH)* [30].

7.5.2 Digital mask programmable hologram

In contrast to the classical digital Fresnel hologram, a BMPH comprises a static, high resolution binary grating, $G(x, y)$, which is overlaid with a lower resolution binary mask, $M(x, y)$. The reconstructed image of the BMPH can be programmed to approximate a target image (including both intensity and depth information) by configuring the pattern of the binary mask with a simple genetic algorithm (SGA). As the low resolution binary mask can be realized using less stringent display technology, the method allows the development of simple and economical holographic video display. Figure 7.15 shows the structure of the BMPH. The pattern $G(x, y)$ is a static, high resolution binary diffraction grating where each pixel is either transparent or opaque. A checkerboard pattern is used for $G(x, y)$ so that the frequency of the grating pattern is maximum along both the horizontal and vertical directions to provide uniform diffraction efficiency along the two directions. The pattern $M(x, y)$ is a binary mask pattern that is lower in resolution than the grating, and is evenly partitioned into square blocks each with a size $k \times k$ pixels, where k is an integer that is larger than unity. Within each square block, all the pixels are identical and set to be either transparent or opaque. As such, the resolution of the mask pattern is $(1/k)$th of that of $G(x, y)$ along the horizontal and vertical directions. Superposition of the pair of images results in the BMPH given by

$$B(x, y) = G(x, y)M(x, y). \tag{7.30}$$

For a given $B(x, y)$ as a hologram, we can determine the location where the image is reconstructed using the diffraction theory developed in Chapter 2. With reference to Eq. (1.30), for a given spectrum of the field distribution $\Psi_{p0}(k_x, k_y)$,

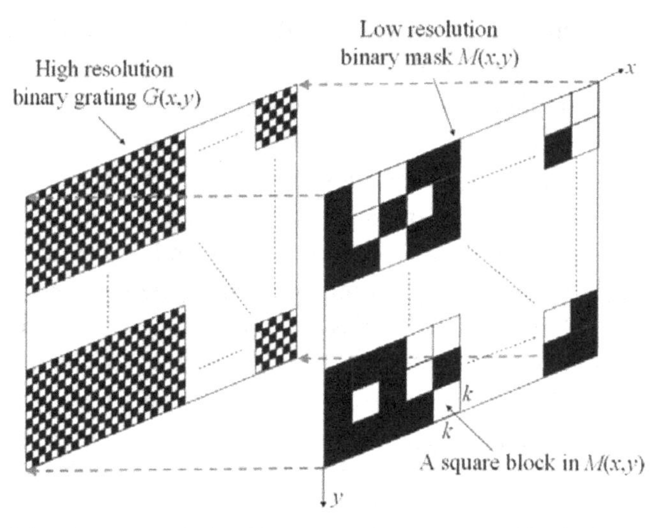

Figure 7.15 Structure of the binary mask programmable hologram, a low resolution binary mask overlaid onto a high resolution binary grating. From Ref. [30], with permission, © OSA.

which is the Fourier transform of the field $\psi_{p0}(x, y)$ at $z = 0$, we write the complex field at z given by

$$\psi_p(x, y; z) = \mathcal{F}^{-1}\{\Psi_{p0}(k_x, k_y)\exp(-jk_{0z}z)\}$$

with $k_{0z} = k_0\sqrt{\left(1 - k_x^2/k_0^2 - k_y^2/k_0^2\right)}$ for waves propagating in the positive z-direction. By substituting $\Psi_{p0}(k_x, k_y) = \mathcal{F}\left\{\psi_{p0}(x, y)\right\}$ into the above equation, we can express $\psi_p(x, y; z)$ as

$$\psi_p(x, y; z) = \psi_{p0}(x, y) * h_r(x, y; z) = \iint_{-\infty}^{\infty} \psi_{p0}(x', y')h_r(x - x', y - y'; z)dx'dy', \quad (7.31)$$

where

$$h_r(x, y; z) = \mathcal{F}^{-1}\left\{\exp\left[-jk_0\sqrt{\left(1 - k_x^2/k_0^2 - k_y^2/k_0^2\right)}\, z\right]\right\}.$$

$h_r(x, y;z)$ is called the *spatial impulse response of propagation of light* [31]. The exact inverse transform has been given but is fairly complicated. For z that is many wavelengths away from $\psi_{p0}(x, y)$ and with the paraxial approximation in the spatial domain, i.e., $x^2 + y^2 \ll z^2$, for any amplitude factors, $h_r(x, y;z)$ becomes

$$h_r(x, y; z) = \frac{jk_0 \exp\left[-jk_0\sqrt{x^2 + y^2 + z^2}\right]}{2\pi z}. \quad (7.32)$$

If we go further using the paraxial approximation in the phase factor above as well, we recover $h(x, y; z)$, the *spatial impulse response in Fourier optics* [see Eq. (1.34), or [31]]. Now, incoporating Eq. (7.31) into Eq. (7.32), we have

$$\psi_p(x, y; z) = \frac{jk_0}{2\pi z} \iint_{-\infty}^{\infty} \psi_{p0}(x', y') \exp\left[-jk_0\sqrt{(x-x')^2 + (y-y')^2 + z^2}\right] dx'dy', \quad (7.33)$$

which becomes the *Fresnel diffraction formula* in Eq. (1.35) when $x^2 + y^2 \ll z^2$ is used in the phase term for the approximation. We shall use Eq. (7.33) to formulate the solution for the BMPH.

When the BMPH hologram, $B(x, y)$, is illuminated with an on-axis planar coherent beam, the magnitude of the reconstructed image at distance z_d can be expressed as

$$I_d(x, y) = \left| \sum_{p=0}^{X-1} \sum_{q=0}^{Y-1} B(p, q) \exp\left[-jk_0\sqrt{[(x-p)\delta d]^2 + [(y-q)\delta d]^2 + z_d^2}\right] \right|, \quad (7.34)$$

where $X \times \delta d$ and $Y \times \delta d$ are the horizontal and vertical extents of the hologram, respectively, as δd is the width and the height of a pixel in $B(x, y)$. Note that the above equation in double summations is simply a discrete version of Eq. (7.33), where we have neglected any constants in front of the integration. Without loss of generality, we assume that the hologram and the image scene have identical horizontal and vertical extents. From Eqs. (7.30) and (7.34), we can deduce that the reconstructed image is dependent on the binary mask pattern, $M(x, y)$. However, given $I_d(x, y)$ there is no explicit inverse formulation to compute $M(x, y)$. In view of this, the inverse problem has been treated as an optimization process to determine the mask pattern that best approximates the target reconstructed image. To begin with, an objective function O_d is defined to determine the root-mean-square error (RMSE) [see Eq. (7.15)] between the reconstructed image $I_d(x, y)$ and a planar target image $T_d(x, y)$ which is located at a distance z_d from the hologram such that

$$O_d = \sqrt{\frac{1}{XY} \sum_{p=0}^{X-1} \sum_{q=0}^{Y-1} [T_d(p, q) - I_d(p, q)]^2}. \quad (7.35)$$

The goal is to determine $M(x, y)$ so that the value of O_d is minimized. It is important to point out that this is different from determining a binary hologram $B(x, y)$ that matches the binary hologram of a given target image $T_d(x, y)$. If this is

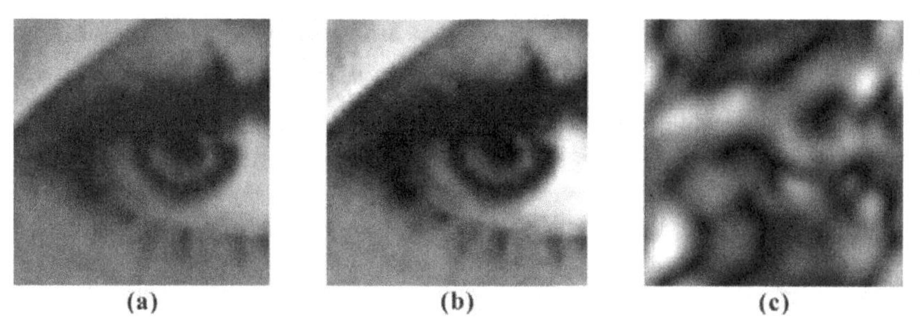

Figure 7.16 (a) A planar image placed at 0.4 m from the hologram. (b) Simulated reconstructed image of the hologram display by a SLM of pixel size 5 μm at 0.4 m. (c) Same as (b) but with SLM of pixel size 20 μm. The number of hologram samples is 256 × 256 and the wavelength of light used is 0.65 μm. From Ref. [30], with permission, © OSA.

the case, the resolution of $B(x, y)$ will have to be identical to that of the high resolution grating. In this method, a low resolution mask $M(x, y)$ is determined so that when it is coupled with the grating, it will result in a reconstructed image that is similar to the target. This is a formidable problem that cannot be solved with brute-force means (such as blind search), as there are $2^{XY/k^2}$ combinations on the mask pattern that can be represented in $M(x, y)$. For example, for a modest square hologram with X and Y both equal to 256, and $k = 4$, the total number of patterns that can be generated is 2^{4096}. In view of this, a simple genetic algorithm (SGA) has been employed [32], a method that mimics the evolutionary mechanism in biological species, to determine the optimal mask pattern. Past research has demonstrated that the SGA is effective in solving complicated optimization problems in many engineering applications [33]. As the principles and details of the SGA have been described in the literature, we shall only present results here. The effect of the use of a low resolution SLM for holographic display is shown in Fig. 7.16. Obviously, the low resolution SLM with 20 μm pixel size has problems reconstructing the hologram as shown in Fig. 7.16(c), whereas the SLM with 5 μm pixel size can display the hologram reasonably well.

When the BMPH is generated for the target image shown in Fig. 7.16(a) with grating size 256 × 256 pixels and pixel size 5 μm square, and a binary mask of 64 × 64 pixels (shown in Fig. 7.17(a)) with pixel size 20 μm, which is four times worse than that of the Fresnel hologram as $k = 4$, the reconstructed image is shown in Fig. 7.17(b).

The main feature of a MPBH is that the target image it represents can be composed by simply changing the pattern on the binary mask. As the binary mask can be implemented with less stringent electronic devices, the method can be used for holographic display where high resolution SLMs are simply not available.

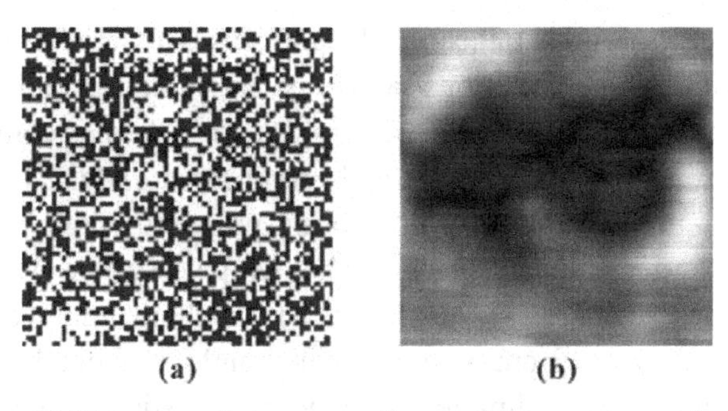

Figure 7.17 (a) Binary mask corresponding to the target image in Fig. 7.16(a). (b) Reconstructed image of the BMPH at 0.4 m. From Ref. [30], with permission, © OSA.

7.5.3 Real-time display

For a given spatial resolution of a SLM, f_0, and according to the Nyquist sampling theorem, the sampling interval of the displayed hologram is $\Delta x = 1/2f_0$. In terms of the *NA*, we have $\Delta x = \lambda_0/2NA$. Assuming the size of the SLM is $L \times L$, the number of samples or resolvable pixels for full parallax is therefore

$$N_{Full\text{-}p} = \left(\frac{L}{\Delta x}\right)^2 = \left(\frac{L \times 2NA}{\lambda_0}\right)^2. \tag{7.36}$$

For a full parallax $L \times L = 30\text{ mm} \times 30\text{ mm}$ on-axis hologram presented on the SLM, and a viewing angle of $\theta = 60°$, the required number of resolvable pixels is about 2.25×10^9 for $\lambda_0 = 0.6328\ \mu\text{m}$. So we are talking about a 2.25 gigapixel SLM and, to put things into perspective, some of the best modern CCD cameras are about 22 megapixels, that is about two orders of magnitude difference between the two devices in terms of the required numbers of pixels. Indeed the required number of resolvable pixels is well beyond the current capabilities of SLMs. Now if we want to update such a SLM with 8-bit resolution at 30 frames per second, a serial data rate of 2.25 giga samples/frame $\times$ 8 bits/sample $\times$ 30 frames/second $=$ 0.54 terabits/second is required and even with state-of-the-art 40 gigabit Ethernet connectivity, it falls short of what we need for the transfer rate. Holographic information reduction has thus become one of the most important topics in real-time holographic display research.

Horizontal parallax-only (HPO) holography is an excellent way to reduce holographic information [34, 35]. The basic idea is that since we are used to looking at the world with our two eyes more or less on a horizontal level, we are usually satisfied with only horizontal parallax. Hence for 256 vertical lines, the number of pixels required is

$$N_{HPO} = 256 \times \left(\frac{L}{\Delta x}\right) \approx 12.1 \text{ million}, \tag{7.37}$$

compared to the case of full parallax of 2.4 gigapixels. The serial data rate reduces to 12.1 million samples/frame $\times$ 8 bits/sample $\times$ 30 frames/second $=$ 2.9 gigabits/second, which is quite manageable with a modern Ethernet connection. Hence HPO holography aims to record and reconstruct HPO information and many HPO holographic systems have been proposed and studied [34, 35]. The first computer-generated HPO digital holographic system was proposed and constructed by St. Hilaire *et al.* [36]. The authors achieved a displayed image of 5 cm $\times$ 3 cm $\times$ 3 cm with a viewing angle of about 15°. However, the first digital holographic recording technique to be investigated regarding the HPO approach was that proposed by Poon *et al.* [37]. The HPO digital holographic system is based on the principle of optical scanning holography. The idea is fairly simple. Instead of scanning a three-dimensional object with a full two-dimensional time-dependent Fresnel zone plate [see Eq. (5.46)] to obtain holographic information, a one-dimensional time-dependent Fresnel zone plate is used to generate HPO holograms. However, these slit-type holograms for each point of the object produce vertical spreading and require the use of a cylindrical lens to compensate for this spreading during reconstruction. Recently, a computer algorithm has been proposed to convert a full parallax hologram to an off-axis HPO hologram for three-dimensional display [38], and three-dimensional display has been achieved [39]. However, the display is confined to simple point objects.

7.5.4 Lack of SLMs capable of displaying a complex function

The reconstruction of the complex hologram is free of the zeroth-order light and the twin image [40]. Unfortunately, modern SLMs can modulate either the phase or the amplitude of incident light, but not both [41, 42]. A brute-force method is to use two SLMs to display a complex hologram by cascading the SLMs, one for amplitude modulation and the other for phase modulation [43–45]. Alternatively, a beam splitter can be used to combine the beams from two SLMs [see Eq. (5.52)], one for displaying the real part and the other the imaginary part of the complex hologram [46–48]. Recently, a method has been proposed to display a complex hologram using a single intensity-modulated hologram [10]. The method involves a standard coherent image processing system with a grating at the Fourier plane. Two position-shifted amplitude holograms displayed on a single SLM can be coupled via the grating and automatically overlapped at the output plane of the image processing system to synthesize a complex hologram. The processing system is shown in Fig. 7.18.

The two position-shifted holograms, $H_r(x - d, y_0)$ and $H_i(y + d, y_0)$ separated by a distance $2d$, are displayed on the x_0–y_0 plane, where $H_r(x,y)$ and $H_i(x,y)$ are the real part and imaginary part of a complex hologram $H_c(x,y)$, respectively, given by

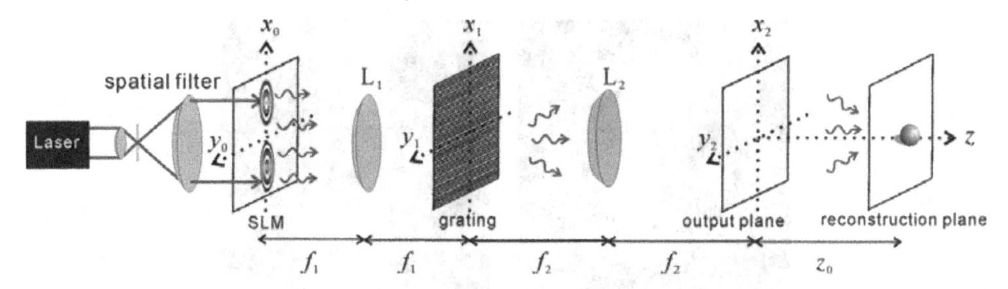

Figure 7.18 Display of a complex hologram using a single SLM.

$$H_r(x, y) = \mathrm{Re}\{H_c\} + I_0, \quad \text{and} \quad H_i(x, y) = \mathrm{Im}\{H_c\} + I_0, \tag{7.38}$$

where I_0 is a positive constant which biases the values of the holograms to be positive so that they can be displayed on an amplitude SLM. The grating employed is an amplitude grating given by

$$g(x, y) = \frac{1}{2} + \frac{m}{2} \cos\left(2\pi \frac{x}{\Lambda}\right), \tag{7.39}$$

where m is the modulation depth of the grating and Λ is the period of the grating. To ensure perfect registration of the two holograms on the output plane, $\Lambda = \lambda_0 f_1 / d$. Also, in order to eliminate I_0 from being displayed on the output plane, a neutral density filter with a phase plate given by $H_{dc} = mI_0\, e^{j5\pi/4}$ at the center of the input plane is used. The overall pattern that is displayed on the SLM is shown in Fig. 7.19(a). Figure 7.19(b) is the display of the output plane where the central portion of the area, which is marked with a dashed square, is the synthesized complex hologram. Figure 7.19(c) is the intensity of the reconstructed image at the reconstruction plane. The simulation results in Fig. 7.19 used gray tone holograms. However, binary holograms can be produced swiftly with a printer. In addition, with binary holograms, we can enhance the storage capacity of digital holograms and facilitate more efficient transmission of holograms.

To accomplish the binarization process, we first multiply the object pattern, u, by a random phase and then calculate the corresponding diffraction field by convolving with h, the spatial impulse response in Fourier optics, to reach the recording plane. The binary holograms are then obtained by binarizing the real and the imaginary parts of the complex hologram by sign binarization. The two binary holograms are, therefore, given by

$$B_r(x, y) = B^0\left\{\mathrm{Re}\left\{\left(u \times e^{j\theta}\right) * h\right\}\right\}, \tag{7.40a}$$

$$B_i(x, y) = B^0\left\{\mathrm{Im}\left\{\left(u \times e^{j\theta}\right) * h\right\}\right\}, \tag{7.40b}$$

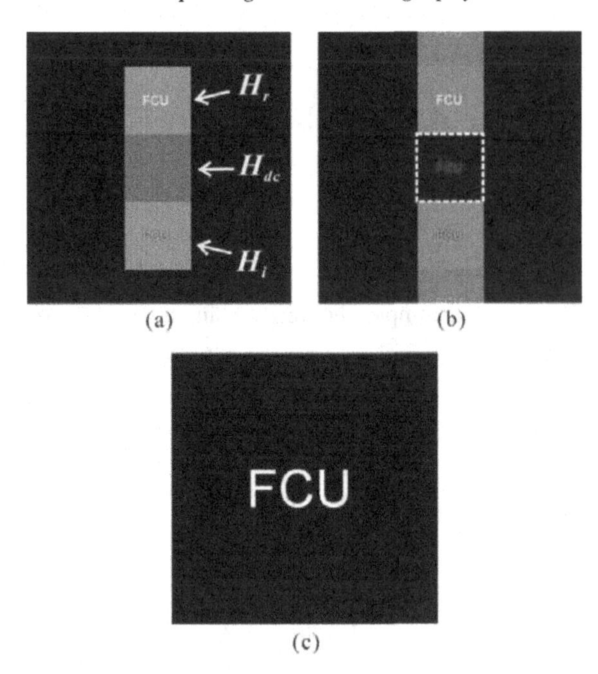

Figure 7.19 (a) Designed input pattern at the SLM, (b) optical field at the output plane, and (c) intensity of the reconstructed image at the reconstruction plane. From Ref. [10], with permission, © OSA.

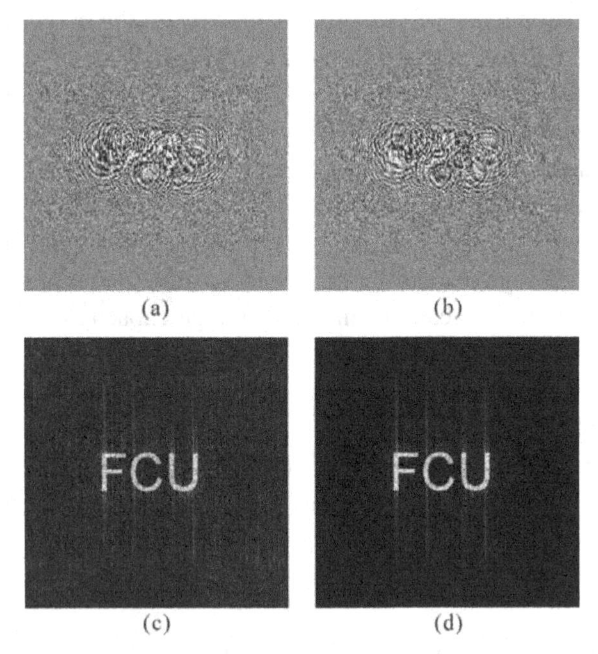

Figure 7.20 (a) Binary hologram $B_r(x, y)$, (b) binary hologram $B_i(x, y)$, (c) image reconstruction using $B_r(x, y)$, and (d) image reconstruction using the complex hologram, $B_r(x, y) + jB_i(x, y)$. From Ref. [10], with permission, © OSA. See Table 7.4 for the MATLAB code.

Table 7.4 *MATLAB code for generating two binary holograms to synthesize a display hologram, see Fig. 7.20*

```
clear all, close all;
I1=imread('FCU500.bmp');%load 8bit,500x500 pixels image
I1=double(I1);
delta=0.01266;    %sampling distance 12.66 um
lambda=0.000633; %wavelength 0.633 um
% add a random phase to the object
Rr=rand(500);
Rr=exp(1i*2*pi.*Rr);
I0=I1.*Rr;
figure; imshow(mat2gray(abs(I0)));
title('original object')
axis off
[M N]=size(I0);
z=-80+lambda/4; %(mm, distance)
r=1:M;
c=1:N;
[C, R]=meshgrid(c, r);
p=exp(-1i*2*pi*z.*((1/lambda)^2-(1/M/delta)^2.*...
     (C-N/2-1).^2-(1/N/delta)^2.*(R-M/2-1).^2).^0.5);
A0=fftshift(ifft2(fftshift(I0)));
Az=A0.*p;
E=fftshift(fft2(fftshift(Az))); % propagation
%binary hologram
Hr=real(E);
Hr=+(Hr>0);
Hi=imag(E);
Hi=+(Hi>0);

%reconstruction of the binary hologram Hr
Ir=fftshift(fft2(fftshift(conj(p)).*...
    ifft2(fftshift(Hr))));
Ir=Ir.*conj(Ir);
Ir=Ir/(max(max(Ir)));

%reconstruction of the synthetic hologram
Hc=Hr+1j*Hi;
I=fftshift(fft2(fftshift(conj(p)).*...
    ifft2(fftshift(Hc))));
I=I.*conj(I);
I=I./max(max(I));
figure;
imshow(mat2gray(Hr));
title('Binary hologram B_r')
axis off
figure;
```

Table 7.4 (*cont.*)

```
imshow(mat2gray(Hi));
title('Binary hologram B_i')
axis off
figure;
imshow(3*Ir);
title('reconstructed image of B_r')
axis off
figure;
imshow(3*I);
title('Reconstructed image of the synthetic hologram')
axis off
```

where θ is a random phase function between 0 and 2π and $B^0\{\cdot\}$ is the binarization operator with the threshold value set to zero, i.e., if the input value is larger than zero, then the output value is 1, otherwise the output value is zero. The purpose of the random phase is to reduce the edge effect caused by binarization.

The two binary holograms $B_r(x, y)$ and $B_i(x, y)$ for the text "FCU" are shown in Figs. 7.20(a) and (b), respectively. Figure 7.20(c) is the reconstruction of a single binary hologram, while Fig. 7.20(d) shows the reconstruction of the synthesized complex hologram. The MATLAB code is listed in Table 7.4. Although some artifacts due to the binarization process exist, most of the background noise, such as the noise due to the twin image, has been removed, as shown in Fig. 7.20(d). For optical reconstruction of $B_r(x, y)$ and $B_i(x, y)$ using the system shown in Fig 7.18, readers are encouraged to check Ref. [10].

Problems

7.1 Generate a detour-phase CGH in MATLAB by setting $\ell = 1$ in Eq. (7.12). Note that there may now be crosstalk between adjacent cells.

7.2 In developing the detour-phase CGH, we have applied some approximations, and they are

$$\frac{p_{mn}k_0d_x}{2\pi f} \ll 1, \quad \frac{q_{mn}k_0d_y}{2\pi f} \ll 1, \quad p_{mn}\sin\theta \ll \lambda_0,$$

$$\text{and } \frac{k_0}{f}\alpha_{mn}d_x \ll 1.$$

Propose a design strategy (i.e., determine the hologram size, cell size, window size, etc.) so that all the approximations can be satisfied.

7.3 Equation (7.26) is a formula derived under the paraxial approximation. Derive the local fringe frequency of the interferogram formed by a plane wave and a spherical wave without applying the paraxial approximation. Then go on to show that the viewing angle becomes

$$\theta = 2 \times \sin^{-1}[\lambda_0 f_0]$$

instead of the angle derived under the paraxial approximation given by Eq. (7.29). Finally plot θ versus f_0 for f_0 up to 1000 lp/mm at $\lambda_0 = 0.6$ μm.

References

1. B. R. Brown, and A. W. Lohmann, Complex spatial filtering with binary masks, *Applied Optics* **5**, 967–969 (1966).
2. A. W. Lohmann, and D. P. Paris, Binary Fraunhofer holograms, generated by computer, *Applied Optics* **6**, 1739–1748 (1967).
3. L. B. Lesem, P. M. Hirsch, and J. A. Jordan, The kinoform: a new wavefront reconstruction device, *IBM Journal of Research and Development* **13**, 150–155 (1969).
4. N. C. Gallagher, and B. Liu, Method for computing kinoforms that reduces image reconstruction error, *Applied Optics* **12**, 2328–2335 (1973).
5. B. Liu, and N. C. Gallagher, Convergence of a spectrum shaping algorithm, *Applied Optics* **13**, 2470–2471 (1974).
6. P. M. Hirsch, J. J. A. Jordan, and L. B. Lesem, Method of making an object-dependent diffuser, *U.S. patent 3,619,022* (1971).
7. R. W. Gerchberg, and W. O. Saxton, A practical algorithm for the determination of phase from image and diffraction plane pictures, *Optik* **35**, 237–246 (1972).
8. J. R. Fienup, Phase retrieval algorithms: a comparison, *Applied Optics* **21**, 2758–2769 (1982).
9. J. R. Fienup, Phase retrieval algorithms: a personal tour [Invited], *Applied Optics* **52**, 45–56 (2013).
10. J.-P. Liu, W.-Y. Hsieh, T.-C. Poon, and P. Tsang, Complex Fresnel hologram display using a single SLM, *Applied Optics* **50**, H128–H135 (2011).
11. L. C. Ferri, Visualization of 3D information with digital holography using laser printers, *Computer & Graphics* **25**, 309–321 (2001).
12. H. Yoshikawa, and M. Tachinami, Development of direct fringe printer for computer-generated holograms, *Proceedings SPIE* **5742**, 259–266 (2005).
13. T. Shimobaba, N. Masuda, and T. Ito, Simple and fast calculation algorithm for computer-generated hologram with wavefront recording plane, *Optics Letters* **34**, 3133–3135 (2009).
14. T. Shimobaba, H. Nakayama, N. Masuda, and T. Ito, Rapid calculation algorithm of Fresnel computer-generated-hologram using look-up table and wavefront-recording plane methods for three-dimensional display, *Optics Express* **18**, 19504–19509 (2010).
15. T. Yamaguchi, G. Okabe, and H. Yoshikawa, Real-time image plane full-color and full-parallax holographic video display system, *Optical Engineering* **46**, 125801 (2007).

16. P. Tsang, J.-P. Liu, W.-K. Cheung, and T.-C. Poon, Fast generation of Fresnel holograms based on multirate filtering, *Applied Optics* **48**, H23–H30 (2009).
17. S.-C. Kim, J.-H. Kim, and E.-S. Kim, Effective reduction of the novel look-up table memory size based on a relationship between the pixel pitch and reconstruction distance of a computer-generated hologram, *Applied Optics* **50**, 3375–3382 (2011).
18. P. W. M. Tsang, and T.-C. Poon, Review on theory and applications of wavefront recording plane framework in generation and processing of digital holograms, *Chinese Optics Letters* **11**, 010902 (2013).
19. T.-C. Poon, On the fundamentals of optical scanning holography, *American Journal of Physics* **76**, 739–745 (2008).
20. P. Tsang, W. K. Cheung, T.-C. Poon, and C. Zhou, Holographic video at 40 frames per second for 4-million object points, *Optics Express* **19**, 15205–15211 (2011).
21. A. Anand, and S. Vijay Raj, Sectioning of amplitude images in digital holography, *Measurement Science and Technology* **17**, 75–78 (2006).
22. W.-C. Chien, D. S. Dilworth, E. Liu, and E. N. Leith, Synthetic-aperture chirp confocal imaging, *Applied Optics* **45**, 501–510 (2006).
23. T. Kim, Optical sectioning by optical scanning holography and a Wiener filter, *Applied Optics* **45**, 872–879 (2006).
24. X. Zhang, E. Y. Lam, and T.-C. Poon, Reconstruction of sectional images in holography using inverse imaging, *Optics Express* **16**, 17215–17226 (2008).
25. D. J. Brady, K. Choi, D. L. Marks, R. Horisaki, and S. Lim, Compressive holography, *Optics Express* **17**, 13040–13049 (2009).
26. P. W. M. Tsang, T.-C. Poon, and K. W. K. Cheung, Enhancing the pictorial content of digital holograms at 100 frames per second, *Optics Express* **20**, 14183–14188 (2012).
27. T.-C. Poon, Three-dimensional television using optical scanning holography, *Journal of Information Display* **3**, 12–16 (2002).
28. M. Stanley, R. W. Bannister, C. D. Cameron, S. D. Coomber, I. G. Cresswell, J. R. Hughes, V. Hui, P. O. Jackson, K. A. Milham, R. J. Miller, D. A. Payne, J. Quarrel, D. C. Scattergood, A. P. Smith, M. A. G. Smith, D. L. Tipton, P. J. Watson, P. J. Webber, and C. W. Slinger, 100-megapixel computer-generated holographic images from active tiling: a dynamic and scalable electro-optic modulator system, *Proceedings SPIE* **5005**, 247–258 (2003).
29. H.-S. Lee, H. Song, S. Lee, N. Collings, and D. Chu, High resolution spatial light modulator for wide viewing angle holographic 3D display, in *Collaborative Conference on 3D Research (CC3DR)*, (2012), pp. 71–72.
30. P. W. M. Tsang, T.-C. Poon, C. Zhou, and K. W. K. Cheung, Binary mask programmable hologram, *Optics Express* **20**, 26480–26485 (2012).
31. T.-C. Poon, and T. Kim, *Engineering Optics with MATLAB* (World Scientific, Singapore, 2006).
32. D. Goldberg, *Genetic Algorithms in Search, Optimization, and Machine Learning* (Addison-Wesley, Boston, MA, 1989).
33. A. M. Zalzala, and P. J. Fleming, eds., *Genetic Algorithms in Engineering Systems* (Institution of Electrical Engineers, Stevenage, UK, 1997).
34. S. A. Benton, The mathematical optics of white light transmission holograms, in *International Symposium on Display Holography* (Lake Forest College, Lake Forest, July 1982).
35. C. P. Grover, R. A. Lessard, and R. Tremblay, Lensless one-step rainbow holography using a synthesized masking slit, *Applied Optics* **22**, 3300–3304 (1983).

36. P. St. Hilaire, S. A. Benton, and M. Lucente, Synthetic aperture holography: a novel approach to three-dimensional displays, *Journal of the Optical Society of America A* **9**, 1969–1977 (1992).
37. T.-C. Poon, T. Akin, G. Indebetouw, and T. Kim, Horizontal-parallax-only electronic holography, *Optics Express* **13**, 2427–2432 (2005).
38. T. Kim, Y. S. Kim, W. S. Kim, and T.-C. Poon, Algorithm for converting full-parallax holograms to horizontal-parallax-only holograms, *Optics Letters* **34**, 1231–1233 (2009).
39. Y. S. Kim, T. Kim, T.-C. Poon, and J. T. Kim, Three-dimensional display of a horizontal-parallax-only hologram, *Applied Optics* **50**, B81–B87 (2011).
40. T.-C. Poon, T. Kim, G. Indebetouw, B. W. Schilling, M. H. Wu, K. Shinoda, and Y. Suzuki, Twin-image elimination experiments for three-dimensional images in optical scanning holography, *Optics Letters* **25**, 215–217 (2000).
41. B. E. A. Saleh, and K. Lu, Theory and design of the liquid crystal TV as an optical spatial phase modulator, *Optical Engineering* **29**, 240–246 (1990).
42. J. A. Coy, M. Zaldarriaga, D. F. Grosz, and O. E. Martinez, Characterization of a liquid crystal television as a programmable spatial light modulator, *Optical Engineering* **35**, 15–19 (1996).
43. L. G. Neto, D. Roberge, and Y. Sheng, Full-range, continuous, complex modulation by the use of two coupled-mode liquid-crystal televisions, *Applied Optics* **35**, 4567–4576 (1996).
44. R. Tudela, I. Labastida, E. Martin-Badosa, S. Vallmitjana, I. Juvells, and A. Carnicer, A simple method for displaying Fresnel holograms on liquid crystal panels, *Optics Communications* **214**, 107–114 (2002).
45. M.-L. Hsieh, M.-L. Chen, and C.-J. Cheng, Improvement of the complex modulated characteristic of cascaded liquid crystal spatial light modulators by using a novel amplitude compensated technique, *Optical Engineering* **46**, 070501 (2007).
46. R. Tudela, E. Martín-Badosa, I. Labastida, S. Vallmitjana, I. Juvells, and A. Carnicer, Full complex Fresnel holograms displayed on liquid crystal devices, *Journal of Optics A: Pure and Applied Optics* **5**, S189 (2003).
47. R. Tudela, E. Martín-Badosa, I. Labastida, S. Vallmitjana, and A. Carnicer, Wavefront reconstruction by adding modulation capabilities of two liquid crystal devices, *Optical Engineering* **43**, 2650–2657 (2004).
48. S.-G. Kim, B. Lee, and E.-S. Kim, Removal of bias and the conjugate image in incoherent on-axis triangular holography and real-time reconstruction of the complex hologram, *Applied Optics* **36**, 4784–4791 (1997).

Index

Printed in the United States
By Bookmasters